THE RETURN OF WOLVES

To Chris for showing me how to think.
To Karen for showing me how to feel.
And to Leo for pointing the ways forward.

THE RETURN
OF WOLVES

*An Iconic Predator's Struggle
to Survive in the American West*

ELI FRANCOVICH

Timber Press
Portland, Oregon

To Chris for showing me how to think.
To Karen for showing me how to feel.
And to Leo for pointing the ways forward.

Published in 2023 by Timber Press, Inc.,
a subsidiary of Workman Publishing Co., Inc.,
a subsidiary of Hachette Book Group, Inc.
1290 Avenue of the Americas
New York, New York 10104
timberpress.com

Printed in the United States on responsibly sourced paper
Text design by Adrianna Sutton
Jacket design by Tyler Comrie
Text is set in Freight Text Pro, a typeface designed
by Joshua Darden in 2005.

The publisher is not responsible for websites (or their content)
that are not owned by the publisher.

The Hachette Speakers Bureau provides a wide range of authors
for speaking events. To find out more, go to hachettespeakersbureau.com
or email HachetteSpeakers@hbgusa.com.

Library of Congress Cataloging-in-Publication Data

Names: Francovich, Eli, author.
Title: The return of wolves: an iconic predator's struggle to survive in the American west /
  Eli Francovich.
Description: Portland: Timber Press, 2023. | Includes bibliographical references and index.
Identifiers: LCCN 2022028218 (print) | LCCN 2022028219 (ebook) | ISBN 9781643260730
  (hardcover) | ISBN 9781643260150 (ebook)
Subjects: LCSH: Wolves—Reintroduction—Washington (State) | Wolves—Conservation—
  Washington (State) | Wolves—Reintroduction—West (U.S.) | Wolves—Conservation—
  West (U.S.) | Gray wolf—West (U.S.)—History—21st century. | Curry, Daniel.
Classification: LCC QL737.C22 F7125 2023 (print) | LCC QL737.C22 (ebook) | DDC
  599.773097—dc23/eng/20220621
LC record available at https://lccn.loc.gov/2022028218
LC ebook record available at https://lccn.loc.gov/2022028219

A catalog record for this book is also available from the British Library.

CONTENTS

INTRODUCTION

It rains early, the drumming on my tent drowning out the noise of the nearby creek. We've made our camp in the middle of what used to be Diamond City. The central remnant of the town is a massive fifteen-foot fireplace, lush with vegetation and moss. Concrete blocks and deep, stagnant wells now full of trash are scattered through the woods. In the 1920s, Diamond City popped up around the logging interests of the Diamond Match Company, at the time one of the largest match manufacturers in the United States. Based in Ohio, they had a plant here in Washington where they harvested white pines. The plant is long gone, but the pines remain.

Daniel Curry, my lanky guide, eats one meal a day. After he eats, he falls asleep no matter what. He breaks his daily fast late at night, which means a quick breakfast for me, and then we're saddling two of Curry's seven horses. Today's mission will primarily be a crash course in equestrian skills, although there is also a trail we need to clear. The trail follows an abandoned roadbed, heading north into thicker and steeper country where cattle graze. And where cattle graze—at least in much of the western United States—wolves roam.

I'm riding Raven, a jet-black Arabian quarter horse. Raven is a beautiful and powerful animal, and the only horse Curry owns that is purebred. Raven, for his part, knows he's a handsome lad and doesn't take kindly to fools. Unfortunately, when it comes to horses (if not other things), I am a fool.

Like most born to this age, nature is a distant amenity for me. I spend more time than many outside. I like to go hiking, skiing, climbing, hunting, and even occasionally bird watching. But there is no urgency in my relationship with the natural world. No necessity. No deep communion. My times outside are sojourns between the next warm meal and engaging Netflix series. I soon discover that working with a horse requires a different level of commitment.

We ride up the narrow road, sawing through downed trees and hacking at encroaching vegetation. We stop often, and I practice mounting and dismounting Raven. It's hard work, made harder by a sun that's burned through the morning's clouds and left the air heavy with moisture.

Curry describes horses as "one-thousand-pound third graders" and treats them as such, employing a brand of progressive discipline that would make even the most conscientious parents look like capricious fools. He also doesn't use bits. "Do this," he says about his generally kind-hearted approach, "and the horse will learn to trust and respect you."

I try, but I keep getting stuck on the first half of his description, the "one-thousand-pound" detail. This is a big animal, I'm beginning to understand. An animal that could toss me from its back and be gone faster than I could holler out in surprise. Thus, I'm a little lax with the discipline. And it turns out that Raven, like many handsome creatures used to skating by on good looks, doesn't like to work all that hard. So, when he is compelled to carry foolish reporters, he likes to eat.

We take a few steps and then he stops, bends his graceful neck, and nibbles. I try making the little kissing sound that, whenever performed by Curry, snags Raven's attention as if a gun has gone off. "Smooch, smooch," I go.

Raven just keeps on chowing down.

And so it goes for several hours: Raven trundling along slowly, eating often, and occasionally heeding my requests. By my standards, this is a victory. The one-thousand-pound third grader has allowed me to live another day. When we turn around, I breathe easier.

Raven too, is excited to go home and suddenly the lethargic pace he's maintained turns into an excited trot. Curry is leading the way on his horse Griph, and this keeps Raven from bolting altogether. But it's quickly evident that I have lost what little control I had. Heading down one particularly steep hill, Raven guns it, dragging me through low-lying tree branches. I squeeze my eyes shut, the branches whipping my face, and remind myself that I have medical insurance.

"Smooch, smooch," goes Curry, and Raven screeches to a stop, ears attentive, eyes on Curry, the perfect picture of obedience. Curry is not happy. Face scratched, I'm just pleased not to have been thrown to the ground. "Raven needs to listen to you," Curry says. "But he will only listen if he respects you. To respect you, he needs to trust you."

I pull some woody debris from my hair.

For the next hour, we work on discipline. We walk the same

section of trail over and over. I ask Raven to stop. If he doesn't listen to me, which mostly he doesn't, I ask him to back up. Slowly, Raven gets the picture. "This is important," Curry says. "When we're out in the real mountains, on treacherous terrain, Raven must obey. Otherwise, Raven might get hurt."

To our left, the hill drops steeply fifty feet to a raging creek jammed with logs, which makes me wonder what truly treacherous terrain looks like.

"Or you might get hurt," Curry adds.

It's slow, frustrating work, for both Raven and me. Curry, for his part, teaches with an extraordinary amount of patience. When we finally return to camp, Raven can't wait to get the saddle off, get under his tarp, and eat some hay.

Unfortunately for him, that's not the plan. Instead, we will rest for about an hour and then head out again. Curry is giving me a crash course in horsemanship, after all.

Toward the end of our break, during which I eat and Curry paces around the camp, the rain returns. Raven is tethered to the horse trailer with a thick line of rope. I approach him to put his reins on, a job I've by now learned to do somewhat efficiently. First I untie the thicker line and retie it around the back of his neck. Then I start to place the reins on Raven, a thick loop of stout rope that sits on his sensitive snout. He snorts, shakes his pretty head, and backs up.

The thick line tightens on the back of his neck, and he panics, rising up on his hind legs and kicking his feet forward. Curry senses the disturbance from the other side of the trailer and yells, "Get away from him."

No need to say it twice. Raven bucks again and slips his neck from the thick lead line. Free, he starts to run. It would be a beautiful sight if I weren't so worried about those hooves.

Curry wastes no time. He follows Raven, not running but not exactly walking, either. "Come on," he says. "Calm down." Raven roars through camp, briefly becoming tangled in the lines holding up our rain tarp and then ripping himself free again. Curry follows. In the woods there is a large hole, one of those old wells now full of trash. Curry gets in between Raven and this death trap and, using a combination of stomps and smooches, he pushes the fed-up animal away from camp and uphill into thicker brush and out of my sight.

I'm left alone, worrying my mistake will hurt Curry, or Raven, or both of them. I'm soaking wet. My pants are torn, boots soggy. This is not the kind of drama I expected to find when I came looking for wolves.

Amidst the dirge of ecological decline, the return of wolves to the lower forty-eight has been a major chord in a chorus of minors. Their return from near-extinction after decades of focused assault is a remarkable conservation and cultural success story. In 1995 and 1996, biologists released thirty-one wolves into Yellowstone National Park. The canids, which had been relocated from Canada, multiplied and spread. Bolstered by similar releases in Idaho and Wyoming, the population grew. Over the following decades, wolves made their way throughout the West, inhabiting territory and habitat that hadn't seen the apex predator since their local eradication one hundred years earlier.

News of their resurgence provoked a whole spectrum of reactions. For scientists, wolves represented a singular opportunity to observe, in real time, the consequences when a long-absent predator returns to an ecosystem. For activists, the return of wolves was a clarion call for conservation. For some ranchers, hunters, and farmers, wolves became a prime example of government overreach and an attack on their values and way of life. For journalists and artists, here was simply a good story that tapped into primal fears and ancient iconography.

Most of the attention was focused on the wolves living in one of the largest remaining tracts of undeveloped land in the contiguous United States: the Greater Yellowstone Ecosystem. Anchored by the eponymous national park, it encompasses parts of Montana, Idaho, and Wyoming. By landmass—between 20,000 and 35,000 square miles depending on where you draw the lines—the ecosystem is larger than eighty-six countries. What's more, the states bordering this wilderness are lightly populated and politically homogenous, with a combined population of approximately 3.5 million people.

In 2008, when the long-legged lopers reached Washington State, they encountered a landscape sundered largely along geographic lines. West of the Cascade Mountains, the Evergreen State is humid, urban, liberal, and increasingly wealthy. East of the divide, it's dry, rural, and increasingly poor. The political beliefs of nearly eight million Washingtonians can roughly be predicted by where in the state they happen to call home.

The wolves arrived from the east and the north—migrating from Idaho and Canada—and their population remains concentrated in the eastern half of the state. And so, every summer for the past decade and a half, when cattle head out to graze the public lands located primarily in Washington's rural, politically conservative, and poorer counties, wolves will kill some of them. Ranchers cry out, saying their livelihood and culture are under attack. In response, state wildlife managers will take to the air in helicopters and kill wolves.

Environmentalists and wolf advocates protest and file lawsuits, arguing that cows are a nonnative species and ranchers are grazing their cattle on public land for a nominal fee. The threats fly in both directions. Wolf meetings are canceled due to threats of violence. The FBI gets called. Wolves poached. Pelts smuggled to Canada, a bloody FedEx package the fateful clue. A state lawmaker suggests sending an environmental activist a severed wolf tail and testicles.

Wolves incite the kind of passions usually reserved for war and infidelity—passions that highlight deep political and social divides. And the cycle continues.

That passion for wolves—both negative and positive—means that wolves get all sorts of media attention and money. Environmental groups milk outrage over the killings of wolves to assist their fundraising efforts. Cattle producers' associations do the same, buying up billboards in eastern Washington featuring the image of a snarling wolf and a tagline urging folks to call the sheriff. State agencies employ multiple biologists focused on Canis lupus, even while more endangered species—sage grouse, for instance—are lucky to have a single fulltime employee studying them. Some of that focus comes from ignorance. And some of it comes from misplaced passion, or pure and simple greed.

But I believe that the primary tension underlying the Wolf Wars is one that's common to all human-nonhuman relationships: the problem of coexistence. Do we have the will and wisdom to coexist with nonhuman animals? That question is particularly pressing in a state like Washington, one that jams humans and animals together. And while Washington may be unique in the United States now—with its dense human population surrounded by wild animals—it likely won't stay that way.

Consider other less populous Western states like Montana and Idaho. Their populations are booming, with people moving into lands once roamed by bears, grizzlies, bobcats, coyotes, wolverines, fishers, and, yes, wolves.

At the same time, larger, more populous (read: more liberal) states that long ago killed their native predators have been clamoring for a touch of wilderness. Colorado has drawn up plans to reintroduce wolves. Meanwhile, California's ongoing rewilding efforts have led to a healthy and surprisingly urban cougar population, and a pack of gray wolves recently arrived in Plumas County along the Nevada border.

The story is similar abroad. In 2015, wolves returned to the Netherlands for the first time in more than a century. Rewilding efforts across Europe are bringing long-absent species back onto a continent that has been manhandled by humans for thousands of years.

In all of these cases, wild animals are coming into contact with highly altered landscapes that demand adaptation and resilience on their part. And yet, the true burden of coexistence will fall upon humanity's collective shoulders. One hundred, two hundred, or four hundred years ago, the answer would have been simpler: move or kill the wild animals. That is no longer a viable approach, as there is no longer an "elsewhere" to which to move these animals. Ironically, and perhaps tragically, as humans move farther from the world of rain, snow, and wind, our desire for a return to the wild—powered I believe by a deep genetic nostalgia—is renewed. And yet, most of us don't have the faintest idea what that kind of life requires.

In September 2019, nine months before I would send him chasing a horse through the woods, I met Curry on a hot and dusty day high in Washington's Kettle Range, hemmed in by thick walls of young pines and wandering cattle. I was touring wolf country with a rancher, a politician, and a biologist all stuffed into a beat-up truck.

At the time, I'd been covering wolf news for two years for the regional paper. It was one of the many issues I was expected to keep tabs on, but unlike other beats, I never felt I had a handle on the topic. Instead, I felt I was mindlessly repeating talking points from both sides of the debate. When caught in the middle of two opposing viewpoints, the normal recourse for a reporter is to find the perspective that most closely represents the middle ground. For issues of wildlife and ecology, this often means speaking with professional biologists and wildlife managers. The prevailing wisdom is that this kind of reporting helps balance out the extreme views.

So I'd attempted to do just that. And yet, I still felt like I had an inadequate understanding of the debate. My reporting had been charged with being anti-wolf. It had also been maligned as being pro-wolf. "Why don't you do your research?" angry readers screamed at me, saying that I was misrepresenting the current science of wolf ecology. The politician I was now riding with had called me late one Friday, spitting mad, alleging that a recent story I'd published had been utterly divorced from the reality on the ground.

All of this disturbed me. Not the anger or criticism, because for a reporter those are as natural and expected as layoffs and pay cuts, but rather the sneaking suspicion that I was missing some bigger, more interesting and important story. While I dutifully recounted the facts, I couldn't help but wonder if there was a more nuanced, compelling, and challenging tale lurking beneath the surface.

I was experiencing the natural limits of rational and reductive thinking. Facts and the scientific process are invaluable tools, but when it comes to looking at the issues where people, animals, and culture intersect, they often fall short.

This is not unique to the Wolf Wars of northeastern Washington. It is present in any discussion that, intentionally or otherwise, pits questions of science against questions of belief. We see it in the politicization of the coronavirus. Likewise with climate change. Both are issues of fact that have become terribly divisive, largely along lines of belief. And it seemed to me that there was perhaps no better metaphor for this fundamental tension in contemporary American behavior than the story of wolves returning to Washington.

So I decided to line up a tour of wolf country with three folks I hoped could help me shake off some of my confusion about this particular story. A rancher who had lost cattle to wolves. A politician who represented ranchers but also played in the halls of power. And a biologist who worked on the ground and seemed to know wolves and people well.

After three hours with them, it felt like little had changed. Over coffee, each of my companions had reiterated points and perspectives I'd heard before. They all came across as sincere and honest people who were staying true to their beliefs, but each of them was also clearly entrenched in their way of thinking. Then we set off in the truck, taking a series of Forest Service roads winding up into the hills. The rancher recounted times he'd stared down wolves, hand on gun. The biologist talked about the biological complexity of the canids. The politician decried the decisions that urban liberals make, how little they understand rural reality. We stopped occasionally to take photos, or to look at some cows standing by the road, their tails languidly whipping back and forth.

And then, at the tail end of the day, eyes heavy from the heat and the early start, we passed a horse trailer parked in a dusty turnout. Two tents were pitched in the dirt under the blazing sun. Behind them, a banked hill led up to thick trees. A large tarp had been stretched between the thickest pines, and two horses occupied the only shade it provided. We stopped and jumped out of the truck. A lanky man with shortly cropped hair, faded cowboy boots, filthy pants, and one gleaming pistol on each hip approached us. He wanted to know what we were doing. Who we were. Why we were there. This is how I met Daniel Curry.

Later I would learn that Curry is a range rider, a job that requires him to spend most of the year in the woods trying to keep wolves from killing cattle and cattle from wandering into the mouths of wolves. He works for the biologist with whom I was traveling. His days roll with the seasons.

This is rugged country, country choked with pines after decades of fire suppression and clear-cutting. Here, he patrols sections of the 1.1 million-acre Colville National Forest on behalf of ranchers who release their cattle onto the land every spring and demand their safe return in the fall.

He spends weeks at a time in isolation, his only company a menagerie of animals (three dogs and three to four horses), working odd hours, heading into the hills at 9:00 p.m. on some days and 3:00 a.m. on others. Typically on horseback (but sometimes on an ATV), he searches for cows and looks for wolves and tries to disrupt the natural outcome of such meetings. He talks about wolves with evident affection, even wearing a ring embossed with the silhouette of a wolf.

Despite that, his most consistent point of human contact is with the ranchers whose cattle he guards. These are men and women who don't wax poetic about the howls of wolves. Politically and socially conservative for the most part, ranchers generally see the natural world as a God-given resource to be used for the betterment of humans. A worldview anathema to Curry, who speaks fondly of individual animals as "beings" worthy of respect and care, regardless of their utility to Homo sapiens. His best friend, I would find out, is his horse Griph.

I didn't know any of this that dusty afternoon, but I sensed in Curry a good story—the coveted fuel of any journalism—and a level of honesty and an air of open-mindedness that I found refreshing after spending the day listening to the usual talking points. He was guarded, but friendly. "I'm busy now," he said, "but why don't you come back tonight? We can chat more." And so that evening I drove back into the woods.

From Republic, a charming albeit decidedly weathered former gold rush town, Curry's camp was an hour drive away. Belying stereotypes about rural life, Republic collects an astonishing array of people: old hippies clinging to the ghost; young back-to-nature types driving banged-up Subarus through washed-out roads; conservative-minded ranchers, loggers, and miners who have seen their ways of life dry up as natural resource extraction has ended in much of the United States; at least one family of immigrants from India. This family owns one of the town's three motels, as well as one of its three gas stations. While getting gas that evening, I asked the attendant how he felt about wolves. He told me that he liked them because they were good for the environment. Still, he'd never seen one, and admitted that he feared them.

After fueling up I drove back into the hills to meet up with Curry. We talked late into the evening, drinking by the fire. The rancher with whom I'd been riding earlier joined us. Whiskey loosened his tongue and I learned that he'd shot and killed a wolf days before in self-defense. Curry flinched at this revelation but said nothing. This rancher, after all, had agreed to try and live with wolves, a minor miracle in eastern Washington. After several somewhat outrageous hours of drinking and arguing with us, he went on to confess that he kind of liked hearing wolves in the hills. Beneath the nearly full moon, someone pulled out a steel drum, and we howled into the night, hoping for a response.

In the morning I returned to Spokane, the second largest city in Washington, where I live and write for the newspaper. I published a story about Daniel Curry and range riding. Life rolled on, but something stuck with me about Curry's quixotic mission. Perhaps it was the doomed romanticism of all those years spent alone in the woods, away from people, defending an animal that most will never see. I was struck by the immediacy of that work. Of the challenge of trying to balance human interests and need with nonhuman interests. On a more fundamental level, I couldn't stop thinking about how little I knew about living in the natural world, unmediated by iPhones, homes, running water, or any of the myriad conveniences that I use each day.

Which is why, this rainy day in June, I'm watching him chase a spooked horse up a hill. He must be wondering why he'd thought it was a good idea to allow a reporter into his life. I can't help but worry that Raven will be injured or lost. I cycle through the different bad outcomes. Broken legs. Gouged-out eye. Wolf attack. Shot by a confused hunter?

Five minutes later, deep into a spiral of absurdity, Daniel Curry returns, with Raven in tow. My ride is canceled. Curry and Raven are going to spend some time together, he informs me. They need to have a talk. Raven knows he's messed up. As the rain comes down, he stands tethered to the trailer, head hanging. Still waiting on the hay.

They ride off and I'm left at camp. As night settles, I build a fire and start drying my boots. The sky is a roiling mass, rain clouds dominating one half, the pastel colors of a setting sun the other.

Several hours later, Curry returns. As he ties Raven up for the night, he's beaming. A wide-open smile, the kind you don't often see on adult men. He nuzzles Raven and comes to the fire. "What a ride," he says. "I needed that." They rode far and fast. Raven responded to his voice alone, the reins barely needed. "We had a talk," he says, "he'll be better with you tomorrow."

The fire is sputtering. A combination of wet wood and rain. "Do you mind?" Curry asks, gesturing toward the smoky mess.

"Go for it, please," I say.

I don't know what he does, a simple readjustment undoubtedly thermodynamically explainable, but in that moment it's magic. The fire roars to life and Curry leans back in his chair. He takes off his boots and beams into the dusk.

Chapter 1

THE WOLF

LeClerc Creek is clogged with fallen trees, giant root balls exposed to the world. At first, it looks to be the act of an epic (and oddly linear) storm. Or perhaps a crew of renegade beavers who've thrown the conventions of their species to the wind and logged simply for the pure joy of the cut. But eventually it becomes apparent that there are no stumps—nor giant holes excavated by toppling trees—along the banks. These pines appear to have fallen from the sky. Which, in a sense, is precisely what happened. A year prior, helicopters flew in more than six hundred trees (dirt encrusted root balls attached), dangling them from cables and laying them across the creek in hopes of providing cover for fish, in particular bull trout.

Curry points out to me that this is a decidedly human innovation: using machines and gasoline to mimic the behavior of beavers. But in an area that has been used so hard, perhaps this kind of intensive intervention is the only way to restore the habitat.

After being dammed in the early 1900s, LeClerc Creek powered a mill for decades. Over time sediment accumulated, narrowing the stream channel, pushing the water's frothing power down into the dirt. As the creek bed deepened, the banks grew higher and more unstable and the water moved faster. Meanwhile, the bull trout, a handsome freshwater member of the Salmonidae family native to the Pacific Northwest, lost a safe place to spawn.

The idea behind these six hundred uprooted trees is to cool the temperature of the creek and slow it down, offering the trout a place to lay their eggs and hide from predators. Not everyone is happy with this plan.

Curry's horse trailer is parked at the end of a road along the creek. He's strung several large raggedy tarps between the trees and the trailer that chatter in the wind and rain. But the weather has taken a turn for the better up here in northeastern Washington. It's going to be a warm June day—a welcome turn of events. As a range rider, Curry spends long days in the saddle searching for wolf sign. Looking for cattle. Learning the lay of the land and trying to interrupt the most natural kind of violence: predator versus prey.

As much as his job involves tracking wolves, herding cattle, and surveying the land, it's also about placating people. We've been here for four days. Earlier in the week the roads and woods were quiet, but with the weekend now in full swing, folks have flooded nearby campgrounds, accompanied by a crescendoing cacophony of braat braats warning of approaching all-terrain vehicles.

The noise starts early at LeClerc Creek. By eight o'clock in the morning, several off-road vehicles have already ripped down the road, their noisy approach sending the Lads, Curry's three lean Dobermans, into a frenzy of protective barking. Fed up with these constant interruptions, Curry constructs a makeshift gate between two trees out of a string with red flagging, which is not exactly legal as we're camped on public land.

It turns most adventurers away, but this flimsy excuse of a gate is no match for Lance. He's the leader of a pack of motorheads—the obvious silvery-haired alpha with a goatee to match—and when he notices the gate, he dismounts and stomps his way down the road wanting to know what in tarnation is going on.

Curry leaps up, intercepting him about one hundred yards from the horse trailer. Like his dogs, Curry is protective of his space, and explains as much to Lance. Lance points out that it's not legal to close Forest Service roads.

"Yeah, but who's going to stop me," Curry says, patting the two guns holstered at his hip.

This aggressive gesture unexpectedly placates Lance. Taking a swig from one of his two beers, he begins to complain about how the Forest Service totally messed up the stream. "It used to be so nice, it ran fast and clear," he says. "Now it's full of logs."

Curry nods, noncommittedly.

"We live up the road," Lance says, "and we're seeing more and more people squatting on the land. Fleeing the cities like rats from a ship, fleeing COVID-19."

"Squatting, really?" I ask.

"Yeah, with their tents. We have to keep an eye on our land. How long are y'all here again?" he says, gesturing at the remains of Diamond City.

So much has changed around here over the past century. Matchmaking used to be big business. Americans used them every day to light wood stoves, cigarettes, gas lights. Now, there is little left of Diamond City, or the US match industry for that matter. Instead, a log-choked stream, spindly and thick new growth forest, old masonry, hundreds of vicious-looking nails strewn everywhere, and a region with more than 15 percent of the population living in poverty.

One week earlier, I picked up a copy of the regional weekly paper, *Selkirk Sun*, from a dimly lit gas station on the edge of the impossibly wide and blue Pend Oreille River. The *Sun* is a small publication that's full of useful reminders about upcoming roadwork in Pend Oreille County along with admonishments from city officials to stop flushing baby wipes. "For sanitary reasons," one article suggested, "a special container next to your toilet can be used for the gloves and wipes—a coffee can with its lid works well."

Two other articles caught my eye. The first was a profile of John Gentle, a man running for county commissioner. Gentle had grown up in the area but left after high school and moved to the coast. When he returned to his hometown, he described the experience as feeling "like a slap in the face." He pointed specifically to the loss of jobs. "The culture had changed," Gentle explained, "the dependency and the brokenness were obvious."

Although the Diamond City Match Company left the area long ago, it was succeeded by many other natural resource-based industries: lumber, mining, manufacturing. "We are very good at processing natural resources," Gentle went on. "We have so much to offer: the rock-bottom living costs, outstanding outdoor recreation, a capable and trainable workforce stuck without much opportunity and optionality for them." Later that year, he would win the election.

The other piece that caught my eye was a letter to the editor defending the Liberty State Movement, a long-simmering desire by some to split Washington State into two states. Leave the liberals on the coast, the argument goes, and let us govern ourselves.

This particular editorial suggested that the United States was in a "similar stupor as Théoden, King of Rohan, in the *Lord of the Rings* Trilogy," reminding the powers that be that there are "millions of combat experienced experts in insurgency" living within our nation's borders. "So before we get pushed too far and things get ugly," the author concluded, "Liberty State is a bloodless act of rebellion, not against the rule of law or the US Constitution, but against those who wish to rule over us. It is a warning and a statement. You've gone too far and we're taking back control."

When Lance speaks about protecting "our land" from the urban rats fleeing disease-choked cities, he is channeling this energy of separation. The same kind of angst underpins much of the controversy around wolves in the United States and elsewhere. The state and federal governments determine how the animals are managed, and governmental employees are, by and large, well-educated and well-paid, at least in comparison to rural counties where median incomes are lucky to break $50,000 a year. The Washington Department of Fish and Wildlife, or WDFW, is based in Olympia, which happens to be on the liberal, populous side of the mountains. Decisions made there, no matter how fair-minded the process may actually be, reek of elitism to those living in the eastern half of the state.

This tension is palpable when talking about wolves, and only exacerbated by the backdrop of a global pandemic and the ongoing racial justice protests in major American cities. Every spring, Daniel Curry is convinced that *this* is the year a range war will break out in Washington, and every year he's wrong, although the regular gunfire from our camp neighbors does not inspire confidence.

This cultural divide is where he hopes to make a difference. When Lance keeps complaining about the tree-clogged stream, Curry gently pushes back, agreeing here and questioning there. "What about the bull trout?" he asks.

For much of modern Western so-called "civilized" history, the natural world has been approached as a thing of utility. Something to serve the needs of Homo sapiens. "The fear of you and the dread of you shall be upon every beast of the earth and upon every bird of the heavens, upon everything that creeps on the ground and all the fish of the sea," says the Bible, and the roots of this belief snake all the way back to the invention of what some believe has been the most revolutionary technology: writing.

The development of written language, specifically the Greek alphabet, started our long divorce from the natural world, argues David Abram in his book *The Spell of the Sensuous*. Suddenly we humans were able to preserve meaning in written characters unrelated to any natural object.

Abram argues that the Greek alphabet took the crucial step of disconnecting meaning from objects. Prior to that, his theory goes, meaning and experience constantly unfolded, like unrolling a never-ending rug. Written language, then, allowed you to slice a square out of this great endless unfolding roll, pack it up and examine it at your leisure.

The slow reduction of the natural world continued, century by century. In 1623 Galileo wrote, "The Universe is a grand book which cannot be read until one first learns to comprehend the language in which it was composed, the language of mathematics." Along with Galileo's many other contributions to science, this insight has found its way onto countless classroom walls.

But the reductivism baton wasn't truly picked up by Western culture at large until René Descartes arrived on the scene. A French philosopher, mathematician, and natural scientist, he argued that material reality—animals, plants, trees, rocks, and flesh—were mechanical, reducible, and mathematically describable.

"By apparently purging material reality of subjective experience," writes David Abram, "Galileo cleared the ground and Descartes laid the foundation for the construction of the objective or 'disinterested' sciences, which by their feverish and forceful investigations have yielded so much of the knowledge and so many of the technologies that have today become commonplace in the West."

That disconnect is particularly vivid in Descartes's dim estimation of animals. Although he was willing to acknowledge that they *were* indeed living—he would give them that much credit—he didn't believe that they could feel pain. Instead, their actions were purely instinctual, and at the end of the day, Descartes believed that they were nothing but fleshy little robots.

This view of the world greatly benefited humans. Lifesaving drugs would be difficult to produce without animal testing, and animal testing would be much harder to defend ethically if we did not still carry some vestiges of Descartes's views. Similarly, the material wonders and luxuries we've grown accustomed to could not exist if it were not for the reductive process that allows us to play God with rivers, electrons, and food coloring.

The average human in North America currently lives in a state of luxury that was unimaginable five hundred years ago. A firmament of wonder-technologies reduces our material suffering. But the natural world—the world of wolves, storms, bears, falling rocks, and swift rivers—remains an unforgiving place, one in which it's hard to make a living.

Currently, more than 50 percent of humans live in urban areas. That number is expected to grow to 70 percent by 2050. That's more than 3.5 billion people living in cities or suburbs, far from forests, mountains or, in the most extreme cases, any kind of green space at all. Never have so many people lived so removed from nature. What will this unprecedented shift in living situation mean for human brains? One early indication: increased mental illness and depression, both of which are found at higher rates in urban populations. There are many other factors at play here, but research has demonstrated, over and over, the importance of connecting with nature for our own well-being, so there is presumably some connection.

On the other hand, the resurgence of wild animals in recently abandoned farm and forestland in the United States and Europe has offered a seemingly counterintuitive narrative. America's century-long retreat from rural life has left formerly developed lands vacant, giving animals like wolves a new place to call home. At this moment you are probably living geographically closer to certain species of wild animals than your grandparents did at the same age, and yet your daily interaction with the natural world is almost certainly more abstracted than theirs ever was.

These two factors—increased urbanization leading to a disconnect from nature; growing numbers of certain highly adaptive species in some landscapes—are fundamentally changing how humans understand and interact with wildlife. This is perhaps best exemplified by a 2019 Colorado State University study documenting an ongoing shift from a "traditionalist" view of wildlife to a "mutualist" view. Traditionalists think of wildlife as a resource to be used by humans, while mutualists believe that animals have their own intrinsic value, separate from serving human needs.

"It's a changing world," says the study's lead investigator, Michael Manfredo. "We've gone from a world where we perceived wildlife as something we had control over and should use the way we wish, to a world where we regard animals as human-like, with a certain amount of rights like humans have."

According to the study, "higher income, urbanization, and education at the state level were associated with a higher prevalence of mutualism orientations among state residents." In Washington, 38 percent of respondents were mutualists and 28 percent were traditionalists. These differences roughly aligned with whether the respondents were living in urban or rural areas.

In general, people that embrace mutualism live far from the wild. They tend to be liberal and urban. They live on the coasts. They go to university and design studies and make policy and head to the mountains on the weekends for adventure and play. They worry about the big picture. About a warming climate. About ongoing and cataclysmic human-caused extinctions. Their mission is noble, but their day-to-day reality is often disconnected from the fact that their way of life is made possible by the subjugation of nature.

On the other side—and these are broad generalizations—are the people who live closer to the natural world. They may work with animals, or perhaps their parents did, and they cling to that culture and history. Usually politically conservative, they view the outdoors as a workplace. Fields to be fenced and grazed. Forests to be logged. Deer and elk to be hunted. They do not hate animals or the environment. They are more connected to the everyday goings-on of the birds and trees than the self-described environmentalists. But they often look at these issues through a narrower lens.

Lance swaggers off and Curry and I return to camp. When we arrived several days ago, we were caught in an early June downpour, the roads muddy and sucking at the tires of Curry's truck. The spot is worth the tricky access, though, because there's enough room to park a horse trailer here and it's located near an abandoned road that leads to wilder, steeper country.

Lance's alleged concern for the land doesn't ring true when I begin to consider the state of the campsite when we arrived. Whoever was here before us used foaming insulation as target practice. The bullet-riddled cans had hemorrhaged thick khaki foam, discoloring the shrubs and grass. Nearby trees were so bullet-torn that they sagged, nearly chopped in two by gunfire.

The previous visitors had also left plenty of non-projectile trash. Boxes of soggy fireworks, paper plates, candy wrappers, eggshells, empty bottles of Sunny D, and a disturbing number of Fruity King Mini Sodas—evidence that brings to mind a small child being raised by ecologically debauched parents with an alarming propensity for sugar.

Cleaning this up took about half an hour. Meanwhile, Curry navigated the horse trailer down the bumpy, rock-strewn road.

While it rained, we strung a line between two cottonwoods and hung a tarp over the top, creating a shelter for Curry's horses, Griph and Raven.

Then, once housed, we turned our attention to setting up our own camp spots, dragging downed trees to collect firewood and eventually cooking dinner (for me) and breakfast (for Curry). Chores ate up most of that first day.

Some of Curry's work is funded by the Northeast Washington Wolf-Cattle Collaborative (also known as NEWWCC, unfortunately pronounced "nuke"). Founded in 2018, the nonprofit's goal is to help ranchers and others who depend on the wellbeing of their livestock learn to coexist with wolves.

One of the primary tools in that effort is range riding. The theory behind range riding is simple, even if the practice is not. Wolves by and large will flee from humans. So, sending humans into the hills along with the cattle will disrupt the inevitable interaction between predator and prey. This may sound romantic, tapping into cowboy imagery and the freedom of a distant horizon, but in reality, it's tedious, often boring work.

And while effective range riders spend much of their time working primarily alone in the backwoods, they also invariably become the face of wolf-coexistence, having to explain themselves to skeptics like Lance. All of which is to say that this morning's interaction with Lance, while not directly tied to wolves or cattle, is very much part of Curry's job description.

By midmorning it's muggy and hot, the rain from the past few days evaporating under a heavy sun. This part of Washington is thick with trees, bisected by numerous roads in varying states of disrepair.

We saddle the horses and head toward one such road paralleling LeClerc Creek. The road itself is solid, wide enough for a four-wheeler or a narrow-bodied pickup. But it's littered with downed trees, the inevitable causalities of winter and spring storms, a rhythmic metronome marking time for boreal life. Curry hacks his way through these obstacles, alternating between using a machete and a saw. By his side, his quarter-horse Griph waits as Curry and I heave logs downhill.

Curry's goal is to clear this road. Why? It heads far back into the grazing allotment (a US Forest Service designated area on which ranchers can let their cattle roam after paying a fee) and once clear, will give Curry and other range riders access to some of the farthest flung cattle. Although Curry and his horse could pick their way through this obstacle course, he wants to be able to travel fast, and at night, if needed. This will allow him to appear wraithlike at all hours, scaring wolves away from the cattle he's guarding.

Mosquitoes swarm us as we go about our work. After every successfully cleared section of road, we ride until we come to another obstacle. Most of the day is spent sawing, heaving, and hoeing, not riding. From past experience, Curry knows that these early-season efforts will pay off later in the summer when wolves and cattle start to meet.

Range riders employ a variety of techniques, the most obvious (and in Curry's opinion the most important) being their simple presence on the land. But Curry also carries a shotgun full of rubber pellets that he uses as a noisemaker and, on rare occasions, a nonlethal weapon. At times he uses fladry, colored string or fabric tied to trees or fences that flutters in the breeze and scares wolves away. He also makes use of fox lights, colored lights that flicker on and off and can be attached to fences near where cattle graze.

Before any of these tools can be used, access must be secured. And so we work, sweat dripping down our faces, the horses whisking their tails to clear their bug-covered rumps. The entire effort ends up seeming a bit futile, but not because of the sheer volume of downed trees. Instead, it's the knowledge that roughly one hundred miles to the west, wolves have already gotten a taste for beef and ranchers are calling for their pelts.

Normally, Curry would be in the thick of it. Racing from ranch to ranch, hill to hill, trying to keep the conflict down. But this summer his boss at NEWWCC assigned him this more sedate slice of Washington wilderness, an area where there are fewer cattle, fewer wolves, and infrequent conflicts. The silver lining is that for the first time in the decade or so that he's worked as a range rider, Curry is clearing trail before the cattle are out.

When he first started riding in this area he was overwhelmed by the sheer density of the place. The hills were packed with trees—new and mid-life growth that made traveling in this country difficult. He couldn't imagine finding cattle in this mess, much less wolves. But after pouring over maps and riding the land a bit, he devised a strategy to clear specific roads and trails to improve his access.

In most ways, he is well suited to interact with people like Lance or the rancher who owns these cattle. Curry didn't graduate from college. He is not a biologist. He knows animals. He knows horses. He knows how to build fires and navigate mountains. While he talked to Lance, I saw respect and, perhaps, envy in Lance's gaze. It may be that Curry is the man Lance wishes he could be. For this reason, Curry can, on his best days, act as a bridge between the biologists and bureaucrats and the ranchers and locals, driven as he is to find a middle ground between these different cultures.

It wasn't always like this, of course. For millennia, coexistence wasn't a choice so much as a reality. Humans lived alongside, ran from, killed, and feared animals in equal measure. But more recently in the long scope of history and evolution, humans have gained a level of control and impact over the world that's shifted coexistence from a fact of being to one of convenience. We have the power to destroy wolves, or any number of other species. Choosing not to—moreover, choosing to return them to the landscape—has been a recent development. To understand why the wolves are back in the hills, and to understand Daniel Curry's work, it's necessary to look back.

In late August 1951, a family of picnickers came across an unlikely sight in rural Washington State: a gray wolf caught in a trap. The 125-pound animal, according to accounts from the time, was impressively built, standing three feet tall, broad chested, a relic of a wilder age. No doubt the family was startled.

As for the wolf—which was likely male based on the descriptions of its size, though no gender was reported—it must have struggled mightily against the steel leg trap, set earlier that month by a government employee. What were they trapping? That detail isn't clear, but they certainly didn't expect to catch a wolf. At any rate, the picnickers weren't equipped to deal with this unexpected lunchtime apparition, so they wrangled up some nearby loggers who killed the animal with axes.

Much as they did elsewhere on the continent, Canis lupus once roamed throughout Washington State, inhabiting a dizzying array of environments. Wolves snagged salmon out of Puget Sound near Seattle, tromped through the deep snow of the Cascade Mountains, and wandered the arid plains of eastern Washington. Along with grizzlies, lynx, and caribou, the wolf is a central figure in Native culture, art, and lore.

But this was the first wolf killed in Washington State since 1924, a grim but accurate descriptor of the condition of Canis lupus in the western United States in the mid-twentieth century. At the time, the finding and killing of wolves was so rare that researchers at Washington State University sent the carcass to the Smithsonian Institute in Washington D.C. to verify that it was, in fact, a wolf. It was. And it would be nearly seventy years until another reliable wolf sighting was documented in the Evergreen State.

Roughly six million years ago, the Earth was tipping into a shape that would almost be recognizable to modern humans. The continents were very nearly arranged as they are today, although North and South America were not yet connected, and India had only just begun its long, slow collision with Asia.

Vast tracts of forest and savannas were slowly giving way to oceans of grass and drier steppes as the climate cooled, a precursor to the coming ice ages. These changes were incremental, inching along at the pace of geology, happening in time scales so vast that they defy comprehension.

While the climate changed, the forests retreated, and the grasslands proliferated, two species began their slow rise to evolutionary prominence. In what is now East Africa, some primates began moving toward these newly opened plains of grass. Gradually they abandoned their boreal homes and learned to walk upright. Their brains grew large, forced as they were to hunt, and avoid being hunted, in these wide-open spaces. Although still a long way from Homo sapiens, the foundation had been set, and in this way our predecessors made the adaptive jump that would eventually give way to us.

Meanwhile (at least when measured by the hazy standards of geologic time), woodland foxes in North America were becoming ascendant. Members of the canid family, these foxes had long been overshadowed by their larger and bulkier cousins. Their rise was abetted by the slow decline of a large-bodied four-legged animal aptly dubbed the bone-crushing dog. After thirty-five million years, these large but slow canines known as Borophaginae found themselves besieged from all sides. At that point felids (read cats) had been in North America for about fourteen million years, slowly establishing a foothold using retractable claws and ambush tactics familiar to any housecat owner today.

But venerable and ferocious species like those found in the Borophaginae family didn't succumb to just one threat. During the time that felids were expanding their range, the climate was cooling. Vast forests were giving way to open prairies, which put the larger and slower animals at a disadvantage. The woodland foxes saw their chance. They bulked up and became better runners.

This triple threat of felids, climate change, and fast and buff foxes proved too much for the embattled bone-crushing dogs.

This new epoch belonged to the footloose and adaptive. Like the African primates, these fleet-footed foxes set the stage for the later emergence of wolves.

A new species arose and *Canis leophagus*, a primitive type of wolf and the progenitor of modern gray wolves and coyotes, emerged in what is now the American Southwest. They spread, traveling into Asia across the land bridge connecting the two continents roughly 3.5 million years ago.

As *Canis leophagus* traveled the globe, evolution's tangled tapestry continued to unfurl. Temporal and geographic separation led to divergence, and in North America, *Canis edwardii* evolved into a species midway between a wolf and a coyote, at least in terms of size.

Time passed and *Canis leophagus* diverged yet again. Several offshoots traveled back into North America from Asia, following migrating herds of mammoths, mastodons, and bison (as did humans). And then, about thirty thousand years ago—a blink of an eye in geologic time—gray wolves joined the canid fray in North America. Here, they joined the dire wolf (of *Game of Thrones* fame), coyotes, *Canis edwardii*, and Homo sapiens in the pursuit of North America's megafauna.

Approximately ten thousand years ago, the wolf survived its first potential extinction. As global temperatures changed, North America lost roughly 70 percent of its megafauna. Known as the Pleistocene extinction, this mass death is generally attributed to a changing climate and the impact of human hunting, although not all subscribe to this theory.

Most of the larger North American animals did not make it. Mammoths died. Giant sloths, too. The American tapir, a large, pig-like animal with a dexterous nose that can still be found in modern-day South America, also blinked out. Consequently, many of the major predators of the day also disappeared: short-faced bears, dire wolves, the American lion.

Among them, perhaps, was a specialized subset of Canis lupus.

In 2007, a scientific paper authored by evolutionary biologists at the University of California Los Angeles and the Smithsonian Institute reported that "Late Pleistocene wolves are genetically unique and morphologically distinct" from modern wolves. The Pleistocene wolves specialized at hunting the megafauna of the time. This meant that they had stronger jaws and teeth than the wolves we know. Their teeth fracture rate was high—about 11 percent—and on par with other carnivores of the time (teeth fracture rates indicate increased carcass scavenging, which in turn suggests escalating competition with other carnivores).

These adaptations meant that these wolves were in their own ecomorph. Not a distinct species, but separated by just enough time, distance, and necessity from modern wolves to manifest some noticeable differences in behavior and appearance.

But why does this matter, when there are wolves in North America today? "The gray wolf did not survive unscathed," concluded the 2007 paper. "At least one ecologically distinct form was lost and replaced by a smaller and more generalized form."

Some of these more generalized forms came from Asia, crossing into North America after the Pleistocene Extinction. By doing so, wolves avoided extirpation, otherwise known as local extinction.

"The survival of the species in North America depended on the presence of more generalized forms elsewhere," the evolutionary biologists wrote.

Ten thousand years later, North American wolves faced their next existential threat—humankind.

Chapter 2

CRYING WOLF

Six million years after brave primates and curious foxes stepped out of the forest, and two years before loggers killed a broad-chested wolf in a trap, Curry's father was born. Like the wolf struggling in that trap, the details of his story aren't remarkable in the particulars, but they do take on meaning when considered in a greater context. Examining the broad strokes of Michael Curry's tragic life can be useful in understanding his son.

Michael had nine stepfathers, his mother cycling "through the dudes" as Curry puts it. The last of those nine men had the surname Curry, which is how the Curry name became attached to him.

Michael Curry was drafted for the Vietnam War as a mechanic, and saw combat while fixing vehicles under fire. His stories from this time were vague—dead friends overseas and a homecoming marked by protests and disdain. He returned to the United States psychologically injured, encountering a country swept up by social, political, and scientific upheavals.

Among the many political and cultural fires of the 1960s and the early 1970s, concern about the runaway degradation and elimination of habitats and species took center stage. President Richard Nixon signed landmark environmental protections into law, including the Endangered Species Act and federal environmental review laws. Nixon himself wasn't particularly interested in the environment. Instead, he hoped to woo a younger generation of voters concerned about the well-being of our planet. There is no indication that Michael Curry concerned himself with the environment, although his future wife certainly did.

In 1981, Michael met Laura Roadifer, who already had a son. The two married and soon after had Daniel Curry, who was born in 1982 on the edge of California's Tehachapi Desert.

Curry was exposed to wild animals at a young age. His mother remembers finding rattlesnakes in their yard. But the most dangerous and unpredictable creature in Curry's early life didn't have fangs—it was his father.

"My dad was just a raging alcoholic," he tells me. "He couldn't just have a beer and relax a little. He would beat my mom. And, like, bad. I remember a few times when she came out covered in blood."

The abuse lasted until Curry was seven years old. One day he returned from school and found a home stripped of furniture and a U-Haul truck parked out front, gassed up and ready to roll. "We're going," his mother announced, and they drove north from California to Snoqualmie, Washington to start a new life. They brought along their dog, Honey.

Laura Curry says that her ex-husband did hit her and once gave her a bloody nose. She only remembers him drawing blood once. He never hit the boys, she says.

They were poor. Pancakes, without butter or syrup, were a typical dinner, and to this day Curry avoids eating them. "A guy can only have so many pancakes in his life," he says.

This existence contrasted sharply with his mother's upbringing. Laura's father was a geologist for Mobil Oil, and she grew up in Tripoli on the Mediterranean Sea, her privileged childhood funded via the extraction of black gold. She didn't come to the United States until she was seventeen years old, which was also the first time she saw a TV.

Laura Curry was the black sheep of her family, more interested in animals than in oil. She worked for the Parks Department in Yellowstone for a time, but never fully pursued her innate curiosity about the natural world.

She worked hard to support her two boys, bouncing from low-paying job to low-paying job. Curry's brother Ryan McLean, who is nine years older than him, was often gone at school and with friends, leaving Curry alone with Honey. They commiserated and played together, building a bond that became, in his words, "such a gift." He remembers crying when he learned about the destruction of the Amazon. The fact that humans were, as he puts it, "destroying species before we even knew what they were," broke his heart.

It's common for children to have an affinity and connection to nature, though it rarely persists into adulthood. Curry was different.

He had already decided at the age of five—under the impression that he had just become a man and needed to find a job—that he would dedicate his life to bridging the gap between animals and humans.

Meanwhile, after the rest of the family fled north, Curry's father was bottoming out in California. Finally, two years later, he would follow the family to Washington State, although he remained estranged from his wife. As Curry grew older, he would periodically try and reconnect with his father, but these efforts were infrequent and devastating.

His father died in 2010, and some of Curry's pain has healed over the years. Although he doesn't have many positive memories of the man, he's able to appreciate the few good ones he does have. During those brief interludes, his father taught him how to throw a knife, fish, and build a fire, regardless of the conditions. His father had been intelligent—that kind of raw, vital intelligence so common among people who work with their hands. He built houses and owned a construction business. "Anything with your hands he could just fucking do," Curry recalls, a trait he has clearly inherited.

During this tumultuous upbringing, he continued to find solace in the company of animals. Much of his adult life has been devoted to returning that love, with his work as a range rider representing a natural extension of that mission, of that deep-seated drive and desire to bring order back into the world. Growing up in a constant state of survival, scrabbling for affection—not to mention food— and being confronted with early violence and neglect, Curry can relate to the roughness of the natural world. In this way he is well positioned to stand as a bulwark between cattle and wolves, and an interlocutor between animals and humans.

But Curry is also deeply distrustful of people and institutions. He often feels like he is being cheated—sometimes he's right, often not—and he frequently struggles to navigate the world of bureaucracy and politics. Like the wolf, Curry has become a generalist, a self-taught jack-of-all-trades, toggling between contexts and skillsets, always trying to stay one step ahead of the next threat, whether real or imagined, forged by the harsh teacher that is survival.

Following the evolutionary chokepoint that was the Pleistocene extinction, wolves proliferated throughout the continental United States, inhabiting a dizzying array of habitat. They hunted the windswept plains of Montana, lolled in the deep forests of Minnesota, and picked their way along the seashore, catching fish from the cold waters of the Pacific. They roamed Europe, hunted in India, and survived the Siberian winters. Social and family-focused, they were for a time the most widely distributed land mammal on Earth.

Prolific travelers, their innate physicality puts human athletes to shame. Wolves can run thirty-five mph and jump sixteen feet horizontally. Their jaws are three times as strong as humans'. In cold temperatures, their circulatory systems regulate the flow of blood to their extremities, allowing them to survive negative-forty-degree nights. To stay warm in such conditions, wolves tuck their snouts between their back legs, using their tails to blanket their faces and turn their backs to the wind. Wolf coats have two layers, first a lighter-colored and denser underfur that keeps them warm, like a down jacket. That layer is covered by long guard hairs, which act as a protective shell against wind, rain, and snow.

Consequently, humans have long used wolf fur for clothing and insulation. And lest you think that desiring a wolf pelt was the sole purview of ancient humans, trappers, and European socialites, consider this: in 1972, the Pentagon ordered more than 250,000 parka hoods to be lined with wolf fur, only to cancel the order after conservationists pointed out that, in order to fulfill it, roughly half of the remaining wolves in North America would have been killed.

As impressive as their ability to live in harsh winter conditions may be, the species' versatility is perhaps even more splendid. Wolves in North America once roamed as far south as Mexico City. There are wolves in India, the Middle East, and at one point they were even in the Gobi Desert.

Unlike cougars, wolves aren't typically ambush predators. Instead of sitting and waiting, they chase their prey, targeting the weakest, sickest, oldest, or slowest in a herd. They're selective and careful when hunting. A well-aimed kick from a deer, elk, or moose will shatter a wolf's jaw—a likely death sentence.

Those hunting instincts remain to this day. One story, told to me by an experienced hunter who wishes to stay anonymous, paints an unforgettable picture of wolves on the prowl.

The hunter was in Idaho's Sawtooth Mountains in the late 1990s, just a handful of years after wolves were reintroduced there. Early in the spring, he drove his Ford Bronco deep into the backcountry near the Middle Fork of the Salmon River, one of the most rugged and remote places in the lower forty-eight. Punching through lingering snowbanks, he descended until eventually arriving at an open basin known as a spring gathering spot for elk.

He'd camped there before. A considerate and ethical hunter, he'd heard wolves howling in the past, a sound he welcomed. Early one morning on this particular trip, he headed uphill to a place where he could see the elk herd spread below him. It was not hunting season; he was just hoping to catch sight of the animals. As the sun rose, he spied at least fifty elk. And then, as if on cue, they all looked toward him.

He froze, eyes glued to his binoculars. Had they spotted him? He gradually realized that they were looking slightly to his left and behind him. There, at the edge of the woods, he saw five wolves staring back at the elk.

The wolves moved forward. Simultaneously, the elk bunched together. The lone calf and its mother dropped behind the phalanx of ungulates and started retreating from the wolves while the main elk contingent moved toward the pack.

A standoff. The wolves eyed the elk. The elk eyed the wolves. The human held his breath. Never had he been so close to wolves before. And then, the youngest wolves started to play, rolling at the feet of the lead male and female. The tension dissipated. The wolves disappeared into the woods. Curious, the man left his perch and walked a road that paralleled the woodland edge. He considered what he knew about wolves. Not much. After all, they'd only returned to Idaho a few years earlier.

He rounded a corner and there, standing in the middle of the narrow road, was a black wolf. The man stopped, his lack of wolf-knowledge suddenly concerning. His vision narrowed, focusing in on the wolf in an act of ancient concentration. Some time passed. He did not move. As his field of awareness expanded, he realized there were two wolves behind him and at least one was in the woods to his side.

He was surrounded, and the wolves appeared to be just as startled by the situation as he was. They'd been hunting elk and now this pale, hairless, spindly creature stood before them. Who knows how long they stood there, facing one another?

Along with being faded by the years, the hunter's memory was hazy from the get-go, as anyone who has had a close encounter with wildlife can understand. When confronted by the vital otherness of wild animals, abstract concepts like minutes have a way of merging together. Time and vision contract. Indeed, descriptions of wolf encounters are notoriously unreliable, with narrators unintentionally exaggerating size, length, and all other matters of fact.

After a period of time, the wolves—again in some silent communication—turned from the man and disappeared into the woods. These were not thick woods. It was an open, mature forest with plenty of space between each tree. And yet, the wolves disappeared as quickly as they appeared. The hunter emphasizes this point over and over when he tells his story.

He stood there, shocked. And then the woods exploded. Elk came flying from all directions. The wolves were hunting.

This is a timeless story. Throughout their respective histories, wolves and humans have often competed for the same prey. They even hunted in similar ways: working as a pack to push prey close enough to attack. Neither wolves nor humans were physically strong enough—or fast enough—to do it alone.

And so it makes sense that there is tension between our two species. Great generalists us both, we came through the ringer of evolution concurrently, often fighting to inhabit the same niche. For the most part, the war between wolf and man has been a slow one, but in recent centuries humans have gained ground. People cleared forests, building roads, cities, and walls, killing wolves whenever and however they could.

It must be acknowledged that our historic animosity toward wolves has been justified. A pack of wolves had the power to ruin a shepherd's life. Consider the numerous wolf-related metaphors sprinkled throughout the English-speaking world: "wolf in sheep's clothing," "a wolfish grin," "hungry as a wolf," among many others, all of which reflect our pastoral past, despite the fact that most of us long ago gave up the shepherd's crook.

It wasn't always adversarial. Certain wolves ingratiated themselves with humans, becoming camp followers of sorts. Some researchers believe that this is the origin story of dogs. According to this theory, humans didn't domesticate wild wolves, no—it was the other way around. Perhaps reading the existential writing in the snow, clever wolves decided to join forces with humans. The cuter, friendlier animals did best, eventually paving the way to full domestication.

Native American tribes had varying relationships with wolves, relationships that illustrated the diversity in thought and belief amongst the first inhabitants of the Americas. However, by the nineteenth century, with Europeans having colonized both continents, the long-simmering tug-of-war between humans and wolves nearly came to an end. Propelled by an unprecedented technological revolution, humans found themselves controlling their natural environment to a degree that would have been unimaginable to their recent ancestors. One of the consequences was that wolves were on the ropes.

For Curry, animals are simple, trustworthy, and in most ways, better than people. It's not hard to trace the development of this viewpoint. An abusive and absent father. An overworked mother. An older brother distanced by age. All resulting in Curry being left alone at home with the family dog, Honey.

They'd play together and cuddle together. Unlike the rest of his family, Honey provided him with constant companionship. Over the years, his love for animals grew, and he slowly amassed a collection of reptiles. As a teenager, he worked at a veterinarian's clinic, learning the basics of animal care.

His basic view of animals is perhaps best summed up in one of his infrequent entries in a personal journal. "If your intentions are true animals will hear you," he wrote in 2008. "The question is not how do I communicate with animals the question is how do I drop the walls that I have built up so that I can hear what they are trying to say."

When he was twenty years old, he started working at Wolf Haven International, an eighty-two-acre fenced refuge on the west side of Washington State located just south of Olympia, the state capitol. Curry got his start at the sanctuary by working in maintenance, but soon edged his way into animal care. He was responsible for feeding the wolves, administering medicine, splinting broken bones, putting dying animals out of their misery, and generally immersing himself in all things wolf. Outside of his work at a veterinarian's office in high school, he otherwise had had little experience caring for animals, but he quickly adapted to the challenges of tending to sick and dying wolves. He bonded with the animals and had his fair share of mystical and hard-to-explain experiences.

In 2003 he euthanized a cancer-stricken wolf named Sioux, having carried the seventy-pound animal to a different part of the enclosure to administer the drugs. As the needle pierced the wolf's skin, Sioux's mate Ruby started howling, a long, mournful serenade. Curry has never been able to explain the experience. But it has become integral to his belief in the inherent dignity, intelligence, and value of animals.

In many ways, working at Wolf Haven was perfect for someone like Curry, whose idealism and passion often result in a near frenetic energy. Even now, in his late thirties, he rarely stops moving. When he is forced to sit, he bounces his leg, or stretches his shoulders, or gives up on the entire charade and walks around while continuing the conversation. Office work is the bane of his existence. He likes to work with his hands, and he loves animals. Wolf Haven allowed him the space to do both.

He often wrote about his experiences in *Wolf Tracks*, the organization's magazine, telling stories about his friends, the wolves he looked after. In a 2008 column he went so far as to state:

In my time here at Wolf Haven I have been given the gift of friendship from people and animal alike. I must say that I most treasure the animals that I have had the privilege to call friends. I find a purity in them that is not commonly found among humans. They have become, and will always be, my family.

Consider a particularly persistent and malignant American myth. One that, despite being discredited in a thousand different ways, remains endemic in our collective consciousness. That is, the belief that precolonial North America was a mostly empty land, ready to be filled with European immigrants. This basic assumption underlies all rationales for expansion. "Manifest Destiny" does not exist without this presumed emptiness.

One of the earliest European accounts of the region that would become known as New England comes courtesy of Giovanni da Verrazzano, an Italian mariner tasked by the King of France to discover whether it was possible to sail to Asia via North America's northern seas. In 1523, as he made his way north from the modern-day Carolinas, Verrazzano "observed that the coastline everywhere was 'densely populated,' smoky with Indian bonfires; he could sometimes smell the burning hundreds of miles away."

This is one of the opening anecdotes in *1491: New Revelations of the Americas Before Columbus*, a book published in 2005 by Charles Mann. According to Mann, Verrazzano's observations were not unique. Most early European accounts recall an Eastern Seaboard full of people and their concerns. Archeologists estimate that more than one hundred thousand people lived there at the time, and that this level of population density wasn't unique to the East Coast.

In 1539, the Spanish conquistador Hernando de Soto came to the area now known as Tampa Bay with six hundred men, two hundred horses, and three hundred pigs. Until his death in 1542 from a fever, de Soto killed, raped, and pillaged his way through modern-day Florida, Georgia, North and South Carolina, Tennessee, Alabama, Mississippi, Arkansas, Texas, and Louisiana. Throughout that savagery, he encountered numerous towns and settlements belonging to Native peoples, including the Caddo people.

After his death, no Europeans returned to the area for more than a century until 1682, when the French explorer René-Robert Cavelier, Sieur de La Salle, tromped through the land surrounding the Gulf of Mexico before being killed in a mutiny in modern-day Huntsville, Texas. He retraced much of the Spaniard's path from a century before, and, as Charles Mann explains:

> La Salle passed through the area where de Soto had found cities cheek by jowl. It was deserted—the French didn't see an Indian village for two hundred miles. About fifty settlements existed in this strip of the Mississippi when de Soto showed up, according to Anne Ramenofsky, an archaeologist at the University of New Mexico. By La Salle's time the number had shrunk to perhaps ten, some probably inhabited by recent immigrants.

This wasn't one isolated evaporative case. Instead, it was indicative of a larger, continent-wide holocaust.

In the time between de Soto's and La Salle's expeditions, the population of Caddo people fell from approximately 200,000 to about 8,500—a drop of nearly 96 percent, according to Timothy K. Perttula, an archaeological consultant in Austin, Texas. Then, in the eighteenth century, the population dropped to 1,400. "An equivalent loss today would reduce the population of New York City to 56,000, not enough to fill Yankee Stadium," Mann writes.

The cause of this tragedy? Probably pigs.

"Ambulatory meat lockers" is what Mann calls these pigs. It was common practice to bring pigs on overseas expeditions. After being released into the new ecosystem, they repopulated quickly and typically followed the humans, providing a renewable food supply.

These attributes—hardy, omnivorous, baby-making machines—were also problems in this new environment. Currently there are more than six million feral swine in the United States. These long-lost descendants of centuries-old expeditions have a tendency to rip up streams, trample fences, and occasionally even kill people. Their destructive prowess continues to incite colorful turns of phrase. One researcher refers to the issue as a "pig bomb."

The ecological damage started pretty much the moment pigs wandered off the boats and onto a new continent. Native peoples had few domesticated animals. Europeans, on the other hand, had stables of them, and had already developed immunity to a number of diseases that likely made the jump from animal to human.

Unaccustomed to these zoonotic diseases, it's possible that the vast population decline in the Caddo people between de Soto's and LaSalle's expeditions was due to an outbreak that originated in pigs.

Some researchers estimate that as much as 90 percent of North and South America's precolonial population died. The severity and rapidness of this pandemic is disputed, and there were undoubtedly other factors at play. Critics of this theory note that a disease that kills all of its hosts cannot last for long, and that even a famously deadly plague like the Black Death didn't kill anywhere near 90 percent of the human population. But no one disputes that rampant disease impacted the Indigenous peoples.

As people died, and pigs and horses spread across the continent, an ecological balance that had developed over thousands of years was upended. Native peoples actively managed their respective environments. They conducted controlled burns, killed predators, and managed prey. Early accounts of the Eastern Seaboard recall a parklike setting, with widely spaced trees maintained by regular cycles of forest fire, both man-made and natural.

In later accounts, colonists describe dark imposing woods, overgrown and neglected. Accustomed to a tamed ecology, these new arrivals from Europe looked at the vast North American wilderness with awe, fear, and greed. Little did they know that what they were seeing was more reminiscent of a garden gone feral than some primordial landscape. These wild lands of the so-called "new world" only bolstered the false myth of emptiness in desperate need of conquest.

European colonists relied on domesticated animals, and domesticated animals were easy prey for carnivores.

Wolves, a mostly exterminated predator in much of Europe, became their natural scapegoats. On the Eastern Seaboard, the obsession with wolf removal among colonists began almost immediately; the first documented wolf bounties were in Massachusetts Bay in 1630. Roger Williams, the minister who founded the colony of Rhode Island, referred to wolves as "a fierce, bloodsucking persecutor."

Colonists moved west, trapping and killing animals, driving them from their homes to clear the way for sheep and cattle grazing. Once plentiful herds of buffalo, deer, and elk were decimated by market hunting and the conversion of forestland to farmland. This, in turn, changed wolf behavior. Forced to adjust to life as a secondary predator, they turned to an easier and more abundant food source: cattle and sheep. In reaction, the ranchers became more entrenched in their rationale for killing wolves. By the 1860s, the depopulation of the West's native animals—buffalo, beavers, grizzlies, and wolves—was well underway.

Back in Europe, the war against wolves was winding down. In the United Kingdom, wolves had been exterminated in the 1680s. In Ireland, this happened in 1770, in Denmark, 1772, and by the turn of the eighteenth century they were gone from the Netherlands. In all of these cases, their extermination was preceded by decades of plummeting populations.

The success of the European wolf campaigns may have led to the fervor with which the colonists took to killing wolves. "The animal husbandry practices of the colonists contributed to the conflict between themselves and predators," writes Rick McIntyre in *War Against the Wolf: America's Campaign to Exterminate the Wolf.* "Few animals were fenced in or closely supervised. As they wandered about looking for grass, they were watched by wolves, bears, and mountain lions."

Practices that moved slowly, from hamlet to hamlet, turned into "the policy of our emerging nation," argues McIntyre. "That heritage of conflict and destruction became part of the American way of life."

Dating back to the beginning of the campaign against the wolf in colonial America, the prospect of their elimination has been championed in newspapers, pamphlets, and books. These writings consistently reveal an underlying motif: nature as enemy, as a force in need of taming.

"Wolves are at the doors of civilization," began an American newspaper story published in 1923. Just five years removed from the carnage of World War I, the article documented an increase in wolf numbers in the United States and Europe. "As in other undesirable things," the article continues, "the blame can be laid directly upon the World War which makes 'hunters of wild beasts hunters of men.'" Depopulated towns and a years-long cessation from most hunting had resulted in increased numbers of Canis lupus, along with a corresponding surge in brazenness on the part of the wolves.

During World War I, entrenched German and Russian forces lobbed explosives and poisonous gasses at each other with no thought of what impact this violence might have on wildlife. When winter arrived in 1917, they started to notice wolves dashing onto the field of battle and attacking the wounded. Groups of battling men, suddenly united by a common enemy, killed about fifty of the animals, which were clearly "desperate in their hunger," according to a story that appeared in the *New York Times* in 1917.

But it was not enough. Fresh packs continued to appear, until, as the story goes:

As a last resort, the two adversaries, with the consent of their commanders, entered into negotiations for an armistice and joined forces to overcome the wolf plague. For a short time there was peace. And in no haphazard fashion was the task of vanquishing the mutual foe undertaken. The wolves were gradually rounded up, and eventually several hundred of them were killed. The others fled in all directions, making their escape from carnage the like of which they had never encountered.

The world's bloodiest and most brutal conflict—one marked by sociopathic levels of stubbornness and a willingness to fight and die, en masse, for inches of ground—was apparently paused in the interest of killing wolves.

The connection of wolves to war is not a new phenomenon. Roman soldiers wore wolf pelts in hopes of channeling some of that ferocity in their own battles. Nazi submarines were called "wolf packs" in World War II. In 2018, the *New York Times* ran a story entitled "A Consequence of Ukraine's War: Less Hunting, More Wolf Attacks." The taming of the North American wilderness was its own kind of battle, one that early colonists eagerly and efficiently participated in. An account from 1852 describes the settling of Otsego County in New York. The settlers cleared the great wilderness, cutting trees down to let light in and generally engaging the so-called "battle of life."

The fervor for this kind of work had its roots in religious convictions. In the Protestant worldview, the wilderness—in all its primeval fecundity—represented the devil. Thus, mission number one was to tame it.

In 1900, author Benjamin Corbin explored this line of thinking in *Corbin's Advice, or the Wolf Hunter's Guide*:

> I cannot believe that Providence intended that these rich lands, broad, well-watered, fertile and wavering with abundant pasturage, close by mountains and valleys, filled with gold, and every metal and mineral, should forever be monopolized by wild beasts and savage men. I believe in the survival of the fittest, and hence I have "fit" for it all my life. . . . The wolf is the enemy of civilization and I want to exterminate him.

This antagonistic view was hardly confined to the wolf. By the end of the 1880s, wildlife of all kinds had been hunted to near extinction in North America. Buffalo herds that once numbered in the millions were down to just a few individuals—a precipitous decline that had a devastating impact on wolf behavior. Buffalo had been the primary food source for wolves in the lower forty-eight, and consequently wolves turned to other, riskier cuisines—primarily cattle. Ranchers cried out and the federal government undertook a massive extermination program.

"The pattern and heritage set by the earliest colonists was carried out to its logical conclusion," explains McIntyre in *War Against the Wolf*. "By the 1950s, the wolf population of the lower forty-eight states had dropped from two million to just a few hundred."

This eradication was fueled first by bounties and then trapping. Although wolf pelts were never as valuable as beaver pelts, they gained in value when the beaver population was trapped to the brink in the late 1800s. And, unlike with beavers, wolf trappers could use poison. To that end, wolfers, as they were known, would first kill buffalo, elk, or other animals. Scattering the remains around the landscape, these men would cut fist-sized gashes in the carcasses and pack them with strychnine.

Strychnine is a poison derived from the seeds of *Strychnos nux-vomica*, a tree native to India and Southeast Asia. It was first brought to Europe to kill rats and mice, and was commercially produced in Pennsylvania starting in the 1830s. A neurotoxin, strychnine attacks glycine, a neurotransmitter that helps control muscle movement. Twenty minutes after ingestion, the doomed animal's muscles will convulse spasmodically and continuously until they become so fatigued that they stop, and the animal suffocates. The aftermath is grotesque: arched backs and splayed limbs. While it's less commonly used these days, it can still be found in some rodent poisons.

The poison-filled carcasses of the 1800s were an effective but indiscriminate killer. Wolves certainly ate them. Coyotes, too. Not to mention dogs, crows, and any other animal that was unlucky enough to happen upon the poisoned meat.

Granville Stuart, one of Montana's first cattlemen who became widely known as "Mr. Montana," described the practice of poisoning meat in his book *Forty Years on the Frontier*. During the winter, the wolves depended on eating buffalo. And so, just after the first freeze, Stuart explained, wolfers would kill a buffalo and fill it with poison. The "brave and intrepid men" repeated the procedure, working in a circular pattern, slowly sowing a line of death. "As soon as the wolves ate the poisoned meat they would die and the bodies freeze solid," Stuart wrote. "One poisoned carcass would often kill a hundred or more wolves." Between 1870 and 1877, historian Edward Curnow estimates that one hundred thousand wolves were killed in Montana alone. Countless other victims of this shotgun-like approach also died.

All of this death—and particularly the unintended poisonings—made the wolfer unpopular among Native tribes, Stuart confessed. "These visits to the baits were always attended by much danger from hostile Indians," he wrote, "and at times the danger would be so great the most fearless wolfer dare not venture out and many valuable skins would be lost."

Francis Cullooyah brings a note of grace and elegance to a café deep in the heart of a casino that is otherwise full of pallid and hungover desperados crumpled over faux mahogany tables. At age seventy-six, Cullooyah is an elder with the Kalispel Tribe of Indians, a small, federally recognized tribe in the northeastern corner of Washington State. He grew up on tribal lands, speaking the Kalispel peoples' native tongue, Salish.

Having seen combat, Cullooyah's uncles returned from World War II and the Korean War rattled men. They spiraled into alcoholism. This disease, Cullooyah tells me, was widespread on the reservation. He was not immune. After attending college, he returned home and got a job as a pipefitter. He worked hard and partied harder, a cycle he'd continue for twenty-three years, at which point he gave up alcohol and dove back toward his roots, first working with incarcerated Native Americans, and then teaching Salish while passing along tribal traditions.

As an elder, Cullooyah advises the tribe on a variety of endeavors, helping them integrate new projects into the cultural matrix. The casino in which we meet is run by the Kalispel. The tribe's reservation occupies less than five thousand acres along the Pend Oreille River, a sliver of the territory the Kalispel people traditionally inhabited, but their work, particularly when it comes to habitat and wildlife, extends into the millions of acres that were once their homeland.

Profits from the casino help fund the tribe's diverse and varied efforts, perhaps none more important than the preservation of fish, wildlife, and habitat. While Cullooyah has little to do with the day-to-day running of these programs, he has guided the overall philosophy—setting the tenor and tone for a department typically staffed by non-Native biologists.

Native lore and archaeological evidence demonstrate that Native peoples throughout North America actively managed their environments, dispelling the myth of an untouched wilderness. Among other techniques, they drove buffalo off cliffs, burned forests, and killed predators. Cullooyah's perspective on wolves—and, by extension, that of the natural resources department—doesn't fit neatly into the talking points of those who love wolves nor those who hate them. Unlike most biologists and wildlife managers, Cullooyah talks about animals in a reverent and personal way.

"When we have our winter dances, when we have our ceremonies, we pray for the wolves," he says. "We pray for the cougars. We pray for the bears."

But those prayers don't change the fact that in the Kalispel's ontology, humans are at the center, a belief that clashes with contemporary animal rights-infused views. Reflecting this fact, the Kalispel Natural Resources Department states that its "fundamental challenge is to provide tribal members with an opportunity to engage in the same cultural practices as their ancestors." These traditional practices included hunting deer, elk, and grizzly bears, and killing predators when necessary.

Like other tribes up and down the Columbia River, the Kalispel sustainably managed prey and predator species. Other regional tribes, for instance, had salmon chiefs who determined when and where to fish, and divided the share of fish among different tribes—a process that mirrors modern fishing seasons and limits as established by fishery managers. Along those same lines, the Kalispel killed grizzly bears in their dens during the winter, a form of hunting seen as barbaric by most today. But how many bears were killed was limited, which helped maintain the grizzly population. As one tribal biologist says, "Native people had a superior understanding of the ecology of their settings, and what they needed to do to not only thrive within that ecosystem, but to maintain it."

Cullooyah points out that maintaining those ecosystems has gotten harder as habitat has been annexed—thus the prayers. But prayers come with a responsibility, and managing ecosystems isn't always pretty.

While Francis Cullooyah provides the why of wildlife management—a why rooted in deep historical and cultural context—Ray Entz provides the how.

Entz, a wildlife biologist working for the Kalispel Tribe, keeps his yard in north Spokane tidy and full of native shrubs (among them, ocean spray and red-osier dogwood), a little grass, and some small trees. He tells me that he intentionally didn't move somewhere more rural, preferring instead to live in the city. He didn't want to add one more power line, one more dog yapping, or one more road to the countryside.

He offers me a seat in his backyard, which is as spacious as the front yard is tidy and contained. Entz, who isn't Native, was hired by the Kalispel Tribe in 1993, a year after the tribe started its natural resource office. He was one of four employees at the time. Now there are seventy. And even though wolves represent a small fraction of the tribe's work, they dominate public interest.

"With wolves, people are interested because it's the most divisive wildlife issue on the planet," he says. "You are pro-wolf, or you are anti-wolf.

"People ask, 'are you pro-wolf?' and I say absolutely," he continues. "We are pro-wolf management. Wolves belong on this landscape. They are part of the ecosystem, but they can't be unchecked. We manage everything else on this planet. We manage roads. We manage people. We manage timber. We manage deer and elk populations. We manage everything and we're not going to manage wolves?"

Part of managing wolves, in this view, is killing them—an action intolerable to some, but one that Entz believes is necessary. Wolves are resilient and reproduce quickly, with litter sizes usually ranging from four to six pups a year. Scientists estimate that an individual wolf will kill about twenty deer annually.

"What do we like to do? We hunt," says Entz, referencing the Kalispel people. "What do we hunt? The same things wolves do. So now we're competitors. So how do you manage deer and elk for people if you don't manage the predator systems for people?"

Other tribes in Washington hold similar management views. In 2019, for instance, the Confederated Tribes of the Colville Reservation in northeast Washington removed all wolf hunting limits for tribal members. In 2021, Colville Tribal hunters legally killed twenty-two wolves, even as the species remained endangered under state law.

"Tribal management is different," Entz tells me. "We are trained by a different philosophy. As a natural resource professional, I was trained in school, but my real training didn't really start until I started working for the tribe, and understanding the complexity of the human environment and how connected humans really are. We are part of that larger ecosystem. We can't divorce ourselves from it."

In some circles today, it's popular to blame the widespread destruction of species solely on modern cultures, typically those originating with European colonialism. This is not backed up by a modern understanding of history. For as long as humans have been humans, we've killed other animals, often wantonly. After all, the disappearance of North America's megafauna is at least partially attributed to overhunting by the land's first immigrants. "Humans are ancient veterans of the art of species cleansing," Dan Flores writes in *Coyote America: A Natural and Supernatural History*. Flores goes on to define "species cleansing" as "the act of pushing fellow animals into black hole oblivion."

Change is the only constant in life, according to the axiom, and it's no different for animals and ecosystems. Throughout most of human history these changes were slow, and people had time to adapt to their new realities. However, as the rate of travel and technological know-how has increased, we find ourselves increasingly ill-prepared for these changes. In the 1970s, futurist Alvin Toffler proposed in the book *Future Shock* that the world is changing too much and too quickly for us to adapt to it. Toffler predicted the shift from an industrial society to what he termed a "super industrial society," more commonly called an information society.

Perhaps this is why wolves have occupied so much mental real estate in recent human history. They harken back to a time when things changed more slowly. They remind us of a past often forgotten, whether ignorantly or willfully. For some, this elicits yowls of rage. For others, this represents a different, supposedly better time. For nearly everyone, the contemporary wolf controversy has transcended the bounds of science and entered the realm of beliefs and mores.

Pushing back against these overwrought reactions is L. David Mech, a US Geological Survey research scientist who has studied wolves since 1958. In a 2012 article entitled "Is Science in Danger of Sanctifying the Wolf?" Mech declared that wolves are simply wolves, one species among many.

> The wolf, while at the top of a food chain and a restored member of the world's most famous National Park and a prominent member of others, remains as one more species in a vast complex of creatures interacting the way they always have. It is neither saint nor sinner except to those who want to make it so.

Daniel Curry is a product of this complicated stew. A boy raised in a culture built on the myth of wide-open, empty lands—a culture malnourished and distanced from nature. Here was a boy who grew up idolizing the trope of the so-called "Western man," the strong individual, the cowboy. A boy who in his own life grappled with a gaping hole where his father should have been. A boy who felt a deep sensitivity for the natural world, who had a rare ability to connect with animals.

He worked at Wolf Haven for a decade. And as he worked there, as he became a man, attitudes in the outside world regarding wolves continued to evolve. Fueled by the idealism of the 1960s, and a rising awareness of mounting environmental excesses, Americans started to care about the environment in a way they never had before. Looking around, they began to notice absences. Wolves gone. Bald eagles, the nation's winged mascot, on the brink. Rivers polluted, catching fire, even. The environmental movement was born in this ferment, winning arguably its biggest victory with the signing of the Endangered Species Act by President Nixon in 1973.

Through all of this, wolves clung on. They survived these lean years in Canada and Alaska and in Minnesota. They made occasional furtive forays into Montana, Idaho, and Washington. Then, in the 1990s, things accelerated, with reintroductions in Idaho and Montana planting the seed for the return of the wolf.

In July 2008, a canine was hit and killed by a car near Tumtum, Washington, twenty-four miles northwest of Spokane. Tests revealed that the dead animal was a wolf. Here was the first physical proof that wolves had actually returned to the state. At the same time, the wolf population in the Northern Rockies—a population including eastern Washington's wolves—was delisted as a federal endangered species. This allowed the state to determine how best to manage its predators. In other words, wolves could now be trapped, moved, or killed by state wildlife managers.

From the moment the state took over management of wolves in Washington, the conflict has hardly abated. Since 2011, the Washington Department of Fish and Wildlife, the agency tasked with preserving, protecting, and perpetuating fish and wildlife statewide, has killed more than thirty wolves in response to cattle attacks. Despite the infrequency of these campaigns, and the fact that the overall wolf population has grown every year since 2008, these killings have infuriated the pro-wolf lobby.

The debate has gotten increasingly heated—environmental activists and ranchers have both been threatened. In 2019, a series of statewide wolf meetings were canceled due to threats of violence, which perhaps ironically came from pro-wolf activists.

Meanwhile, Daniel Curry continued to work at Wolf Haven, tending to captive wolves. He read with fury and horror the accounts of state wildlife managers killing wolves in response to wolves killing cattle. He believed that ranchers should not be allowed to let their cattle roam on public land. Wolves had a right to be there. Cows did not. He started to wonder if he might be more useful somewhere else.

He knew that he was preaching to the choir when he wrote about his animal friends in *Wolf Tracks*. On top of that, he was tending to wolves that were already saved, already destined to live out their remaining days in the safety of a pen.

"I'm not really saving any wolves," he says about his frame of mind back then. "I'm not really helping build that bridge that I always felt like I was. I put my finger in the middle of all the wolf packs. And I'm like, 'I'm going there.'"

In 2012, after the state killed an entire wolf pack following repeated attacks on livestock, Curry had had enough. He packed his bags and headed east, into the heart of wolf conflict.

The fact that ranchers and wildlife agencies would have any restraint at all when it comes to killing wolves underscores just how dramatically our view of the natural world has evolved over the past century. Wildlife, once widely feared and demonized as a force of the devil, one to be conquered by plow and sword, is now revered and protected by a bevy of laws.

The American model of wildlife conservation is heralded around the world. It is undoubtedly a success story. But coming so closely on the heels of Western expansion, the sudden-seeming shift in view can be disorienting.

Adding to that disorientation is the undeniable truth that wild animals, and predators in particular, are just that—wild. Living peaceably alongside them is not something that modern cultures are accustomed to doing, particularly now that the old model of trapping, hunting, and exterminating them is no longer acceptable.

Wolves have survived other near extinctions. Their adaptability as generalists, along with their propensity for migrating, saved them back then. But as our world becomes increasingly bifurcated by human endeavors, I do not think that wolves, or any large-bodied species, will survive without buy-in from humans.

It's clear that the old model of coexistence—*you live there, I live here*—won't work. There is no "there" anymore. No here. Our vast reach as a species has made the world one, whether we like it or not. The problems posed today are unique, so we can't necessarily rely on the solutions people have relied on in the past. We need to take a new approach and Curry believes he has an answer, but it's not simple.

Chapter 3

LOOKING
FOR SUPPORT

Before he packs his truck and drives west to meet the governor, Curry buys a new pair of pants, a not insignificant purchase for a man who makes less than thirty thousand dollars a year practicing the world's oldest profession in the West's second-most populous state.

The pants, which required Curry to drive an hour-and-a-half from his log home near the Canadian border to purchase, are a compromise of sorts, conforming to the letter of the law of urban sociability, if not the spirit. And hey—even if the meeting goes badly, he'll still have a new pair of thick canvas work pants.

He's coming off another frustrating season of range riding, one in which wolves killed cattle (again) and in response the state killed wolves (again). Driven by the idealism of true believers everywhere, he remains convinced that there must be a better way, a way in which no wolves die, and no ranchers feel betrayed.

Despite that passion, or perhaps because of it, Curry is broke (again). And now, with winter coming on and the cattle mostly off the land and the wolves hunkered down, Curry knows that it's time to switch gears and consider his future. Which is why he's

---

heading to Seattle to hobnob with wealthy and powerful environmentalists after being asked to talk about range riding.

Governor Jay Inslee is tall and moves with the physicality and confidence of a former high school athlete (he was a star in basketball and football in the late Sixties). In 2019, he launched a presidential bid that focused heavily on the environment. He was seen as the climate change candidate and garnered some headlines for his policies, but ultimately failed to stand out among the other distinguished silver-haired white men in the field. After suspending his campaign, he returned to Washington to defend his seat in the 2020 gubernatorial election. While Inslee's presidential campaign proudly heralded his commendable environmental policy record, the reality is a bit messier back in his home state, where wolf advocates have been criticizing the state's wildlife management agency for killing wolves.

Luckily for Inslee, more than 75 percent of Washington State's population is jammed into the western, wealthy, liberal side of the state, primarily in the Seattle area. In 2016, he handily won a second a term as governor. And while that liberal bastion eases election night doubts, it can make governing a fraught task.

An old idea to create a new state—the State of Liberty—remains an ever-popular vision in some rural Washington and Oregon conservative circles. These secessionist tendencies are not uncommon in the West, particularly in coastal states where political control is centralized in the urban cores, leaving many rural communities feeling disenfranchised and disconnected. For example, some residents of southern Oregon and northern

California have proposed redrawing the borders around a collection of counties and calling this new entity the State of Jefferson.

The schism between urban and rural populations pops up in just about any policy or political discussion. Vehicle emissions. Public transit taxes. Climate change initiatives. Gun control. But wolves have highlighted the cultural and political divide in a way that other issues haven't.

Environmentalists have been lobbying Inslee to intercede on the wolf controversy for years. The pressure only intensified after the state killed nine wolves in 2019. During his presidential campaign some whispered of hypocrisy: *the environmental candidate allows the killing of wolves in his own state, the shame.* Billboards decrying the deaths were erected along the I-5 corridor in Seattle. Eventually, in a move seen by some as a calculated sort of political appeasement, Inslee publicly asked the state wildlife agency to kill fewer wolves—an unusual level of micromanagement by the chief executive of a state containing more than seven million people.

As an idealist who firmly believes that wolves and cows can, for the most part, peacefully coexist, Daniel Curry rails against any compromise that seems to threaten that potential peace. For this reason, he detests politics, which he believes has "butchered" efforts at coexistence. Nevertheless, hard financial realities have forced him to swallow his pride and idealism in the search for financial and political support.

And so, after driving six hundred miles west, he parks his truck at a mansion overlooking Lake Washington. It's the kind of house that has its own name—Harwood—and "is so grand that you hesitate to disturb its pristine silence by ringing the doorbell," according to a *Seattle Times* article from 2007.

It's the domicile of Jennifer McCausland, a rich Australian émigré with a passion for classical music, politics, and wolves. For Curry, who has spent the previous week trying to eat as little as possible in order to save money, the setting couldn't be more foreign. The rich and powerful, including a high-ranking Taiwanese politician, drink wine and eat delicate appetizers, all hoping to bend Governor Inslee's ear.

Even with Curry dressing his best, there is no hiding the gulf between his world in the woods and the mahogany-paneled home in one of the ritziest neighborhoods—where median home prices have reached as high as nine hundred thousand dollars—in one of the nation's richest cities. "I think I would be way more comfortable strapped with T-bones in front of a grizzly bear," he says about his foray into the dangerous world of small talk, wine, and hors d'oeuvres.

When Curry moved east to wolf country in 2012, he drove in the exact opposite direction of today's journey to meet with the governor. Back then, crossing the Cascade Mountains on Interstate 90, Curry caught a glimpse of the other side of Washington. It looked like a completely different state. The climate, for example: as Pacific storms slam into the Cascade Range, they unload their payload—rain or snow. Spent, they cross Washington's flat and dry middle before running into the glacier-carved valleys of the Colville National Forest and the western slopes of Idaho's Rocky Mountains.

Curry brought along his animals, including his horse, Griph, along with three dogs, three cats, a bird, and five reptiles, among them a red-tailed boa. Sight unseen, he rented a house north of the town of Colville, intent on starting an organization that would specialize in nonlethal predator-human interaction. He named his business Guarding the Respective Interests of Predators and Humans, a clunky name with a meaningful acronym, and started showing up at wolf meetings, advertising his services and promoting GRIPH to anyone who would listen. A dedicated herpetologist, Curry cheerfully burned sixteen thousand pounds of wood that winter to keep his home reptilian-friendly.

However, things did not go smoothly. Within a year of moving east, he realized he'd accidentally rented a house owned by members of a Christian Identity church known for its violent and racist ideology. He rented another house, only to have the owner sell it months into his lease. He spent a winter living and working out of his horse trailer on public land before eventually finding a new home north of Colville, the county seat of Stevens County.

In addition to grappling with these housing woes, Curry's mantra of coexistence was being met with skepticism and outright hostility. At a state wolf meeting held in Colville in 2013, he addressed a crowd of more than three hundred Stevens County residents. "I don't want my horses to be eaten by wolves," he said, attempting to reassure them. "I'm working on deterrents. . . . We have to figure out how to coexist."

The crowd groaned and booed, according to a newspaper article that ran after the meeting. After Curry finished his presentation, one man told him, "After hearing you talk, boy, I think you need to grow your hair out and change your clothes. You seem more like a hippie to me."

Despite coming up against this kind of resistance, Curry did make some progress. At that same meeting, a ranching family thanked him for sharing his perspective. Yet he still walked to his truck with his hand on his knife.

In 2017, he stumbled upon a public relations coup. A rancher called. Two cougars were underneath his barn, could Curry get them out? If not, the rancher would kill them. It was ten o'clock at night and Curry didn't hesitate.

The video he took of the encounter is panic inducing. The two cougars are crouching underneath a barn. The oppressive smallness of the space is palpable even in the low-quality recording. Curry inches toward them, the light from his headlamp illuminating the cats' impassive gazes.

"There's the meow meow," he says in the video. "Holy balls, we're close together, aren't we?"

Curry is tall and muscled from outdoor work, and it's a tight squeeze for him, getting into that crawl space. But here he is, five feet from the two big felines. He addresses them calmly. "You're beautiful cats. I appreciate you guys. That's why I'm under this barn, risking my life to get you out."

For the ranchers of northeast Washington, and much of the West, cattle ranching means public land grazing. Every spring, ranchers release their stock onto National Forest grazing allotments, for which they pay a nominal fee. In the fall they round them up again. Most have little contact with their cattle during the intervening months.

———

Despite America's attachment to the image of cowboys on the wide-open range, public land ranching, also known as rangeland ranching, is not nearly as profitable as feedlot-style ranching. According to the magazine *High Country News*, the "heyday" of rangeland ranching ended in the 1880s. Since then, it's represented a relatively small subset of the nation's cattle industry.

When compared to a feedlot, public land grazing is bucolic. Free-range animals are healthier—and arguably happier—than their feedlot brethren.

In a nearly predator-free landscape, this was a tenable state of affairs for ranchers. They had access to thousands of acres of land for which they pay roughly $1.35 for a cow and calf, per month. But wolves have changed the equation. Since 2011, wolves have killed at least ninety-five cattle in Washington and injured an additional sixty-five. On the surface, this is a small number of casualties out of an estimated 230,000 total beef cows in Washington.

However, ranchers use three main arguments to refute the notion that wolves have an insignificant impact on their cattle.

First, these deaths are only the ones that the state will confirm as wolf kills. Many more cattle simply disappear. Others, when their bodies are found, are too decomposed for the authorities to confirm what caused their deaths. Some research indicates that for every dead animal found, three more have been killed, with those remains not found.

Second, the simple presence of wolves on the land can stress out the cattle, leading to more miscarriages and skinnier animals. With a single pound of beef representing roughly $1.50, skinny animals mean skinny paychecks.

Third, wolves represent a tipping point for ranchers. They are already struggling to compete with feedlots. Washington ranchers are geographically isolated, far from any major roads or active rail lines, which means that getting their cattle to feedlots and slaughterhouses is a more difficult and expensive process for them than it is for other cattle producers. The increased mortality rates in their herds due to wolf attacks threaten to send them over the edge.

"You can't take those kinds of losses and stay in business," a rancher named Joel Kretz tells me. Kretz is also a state representative.

The geography of northeast Washington makes these losses seemingly inevitable. The cattle are wandering through steep, forested country. Many of the tried-and-true methods of keeping wolves from cows—pastures, dogs, lights, fladry—don't consistently work in this landscape. Instead, the most effective solution also happens to be the most difficult: human presence. Wolf attacks on humans are rare, and if a pack of wolves senses humans nearby, the pack will flee.

When viewed from a certain context, the return of wolves to Washington is a success. Even with annual losses to poachers, wildlife officials, and as roadkill, the state's wolf population has grown each year, and isn't in danger of going extinct. No matter how you look at it, the wolves aren't single-handedly destroying the ranching industry or endangering humans. Washington's ranching industry already faced difficult odds long before the wolves arrived, what with being located far from slaughterhouses and smaller in size than the larger operations with whom they compete.

But tolerance for wolves is on the ropes, at least in Curry's view. The rhetoric has turned increasingly violent. Every time a wolf kills a cow, or the state kills a wolf, the divide widens. That leaves Curry wondering if the sacrifices he's made—the broken bones, the sleepless nights, the eviscerated bank accounts—have been worth it. Not to mention having to endure the politics and bureaucracy. "I've never been this close to quitting," he says.

He is not entirely joking when he says that he would feel more comfortable wrapped in meat while facing a hungry grizzly than he does while socializing with well-heeled humans. And despite driving all the way to Seattle in hopes of finding support for his work, he is destined to be disappointed.

The governor is late.

Punctuality (or at least some kind of effort resembling punctuality) is important to Daniel Curry. As someone who spends most of his time out of cell service, he expects people to be where they say they're going to be, or at least within spitting distance, at the agreed-upon time. For Curry, there is rarely any last-minute texting to change plans.

This unexpected delay is not easy for him. And the setting is just a little too much. The Gucci shoes. The lush house full of art and mahogany and people dressed in expensive suits and doused in sweet smells. Overwhelmed by the colognes and perfumes, he walks outside three times while waiting.

Finally, the governor arrives, an hour and a half late. The gathered diners are allowed to ask Governor Inslee questions, but Curry is disappointed to discover that the questions have clearly been decided upon in advance. And when he finally gets the chance to talk to Governor Inslee, he realizes that he too is expected to ask softball questions. Softball questions that ignore the underlying urgency of Curry's mission: *How can we learn to live with wolves and by extension, learn to live with one another?*

The entire evening is so disconnected from Curry's reality, from his lived experience on the land, that he is left wondering what the point of the gathering might actually be. It's clear to him that the governor doesn't appreciate the urgency of the situation, and that the others don't have any idea what it's like to live near wild creatures. Ranchers have more of an idea, but they too are disconnected, buffered as they are by technology and decades of working in landscapes nearly free of predators.

On its own, perhaps this would not be not such a big deal. Wolves will survive. Environmental groups will continue to fundraise. Ranching won't fail solely because of wolves. But Curry believes that the Wolf Wars are a symptom of a larger problem: the political and cultural divide in Washington and across the nation.

As far as he can tell, this divide is not only at the core of all wildlife tensions in Washington, it's at the core of the geopolitical tensions in the country as a whole. All too often these sorts of tensions become shouting matches, with each side accusing the other side of ignoring the facts. The Wolf Wars are a case study of this kind of division.

Chapter 4

INTO THE WOODS,
INTO THE WEEDS

On a warm day in early March, the snow lingers in patches along
the sides of the highway, with the intrepid flower breaking through
here and there. Today I'll be tagging along with Ben Maletzke,
Washington's statewide wolf specialist, and Grant Samsill, a wildlife conflict specialist for WDFW. We're heading into the hills
above the Columbia River to count wolves.

We meet in a dusty parking lot behind the US Forest Service's
administrative office in Colville. The county seat is full of log
trucks, federal and state government offices, fast food, and an
ever-growing number of breweries.

I'm riding with Maletzke, who has towed two snowmobiles all
the way from his home in Cle Elum in central Washington. The
five-hour drive required a 1:00 a.m. departure—a not uncommon
schedule for the man tasked with managing wolves throughout
the state.

As we barrel west on Highway 395, the mostly two-lane road
connecting many of Washington's smallest and poorest rural
communities, Maletzke tells me about his goals for the day. The
plan is to drive up and down snow-covered Forest Service roads,

looking for wolf sign, whether footprints or scat. Whatever information Maletzke and Samsill glean will be fed into a report that WDFW releases every April.

That report will guide management decisions for next year, establishing a minimum number of wolves and packs in the state. This makes it a hotly anticipated piece of research (at least by research standards) that occasionally draws criticism from those who believe the wolves have either been under- or overcounted. Maletzke is aware of these complaints but trusts his methodology. In fact, he actually derives some confidence from all the criticism. "When everybody hates you," he says, "you're probably doing it right."

Maletzke is a burly man with closely cropped hair and a rubicund face clearly accustomed to wind, rain, and sun. His soft voice and high-pitched laugh offset his otherwise rugged appearance. For the past decade, he's worked as the statewide wolf specialist. Before that, he studied Canada lynx, another predator native to Washington. He confesses that working with wolves has been the hardest job he's ever had. But he doesn't blame it on the controversial animals. It's the people.

After twenty minutes on the highway, we approach a large body of water—easily mistaken for a lake but actually a dammed river. This testament to humanity's God-like powers happens to be the mighty Columbia. Although the sky is clear, a low-lying bank of fog hugs the river.

Despite being rural and sparsely populated, this particular landscape has been molded by human hands and ambitions for thousands of years. The most visible changes are also the most recent, starting with the Columbia River itself. From its headwaters in the Canadian Rockies, the river flows first north, and then hooks south into the United States, eventually demarcating most of the border between Oregon and Washington before draining into the Pacific Ocean. At more than twelve hundred miles long, it's the fourth largest river in the United States when measured by volume, and has always been important to humans. Indigenous people (currently represented by at least four major tribes) traveled along the river, lived beside it, and depended on the river for salmon, one of their primary sources of food.

Now, hydroelectric dams dot the river—temples of productivity supplying more than 40 percent of the nation's hydroelectricity—and stymie its flow. These dams, the largest of which was finished in 1942, flooded thousands of acres.

Founded in 1891 by a passel of local white businessmen (with the financial backing of a rich New Yorker), the town of Kettle Falls once had resort-town dreams. The area's natural beauty, along with the fantastic salmon fishing, excited their pecuniary fantasies.

Wolves, bears, cougars, and humans have all roamed widely throughout the Northwest, but if there's a single species that can claim historical dominion, it's salmon. Washington's rivers, including the Columbia, seethed with them. Local tribes organized their years around the fish, as did bears, and to a lesser extent wolves. And the area now known as Kettle Falls was the happening spot. Early written accounts recall a river so jam-packed with the big, nutritious fish that one couldn't throw a stick without hitting a Salmonidae.

Every year, at least fourteen tribes would gather at the falls to fish for salmon and trade news and goods. This gathering place was called Shonitkwu, the Salish word meaning noisy water, because the river used to drop more than fifty feet, tumbling over a series of massive quartzite blocks. The frothing, roaring water could be heard from miles around.

When it was constructed in 1942, Grand Coulee dam raised the waters behind it three hundred and eighty feet and flooded more than twenty thousand acres of land. Standing one hundred miles downstream from Kettle Falls, the dam's five-hundred-fifty-foot-tall cement edifice nonetheless silenced the noisy water of Shonitkwu.

The town itself relocated as the waters rose. Now, Kettle Falls is a blip on the road with an unusually good health food store and a name that makes no sense. Look at any map of the western United States and you will see landmarks with similarly out-of-place names: American Falls, Priest Rapids, Celilo Falls.

When the last wolves were driven from Washington in the 1930s, Kettle Falls still made descriptive sense and the Columbia River was not nearly so torpid. Returning after nearly a century's absence, wolves have found a different world—one bisected by roads and trains and bounded by obese rivers, sprawling towns, and hills of cattle.

The wolves of the Wedge Pack live jammed (dare I say wedged?) between all of this. To the east of their territory is the Columbia River. To the west is Highway 395. First confirmed to have returned in 2012, wolves in this area are no strangers to controversy. That same year, the state killed the wolves in the pack after repeated attacks on livestock. However, wolves have continued to repopulate the territory, a reflection of the area's prime habitat and proximity to larger wolf populations in Canada.

At the time of our snowmobile trip, the state estimates that three wolves call this slice of land home. But those numbers are based off data from the spring of 2019, nearly one year ago. Today's mission is to find out what's happened to them over these past nine months.

We drive north and start gaining elevation. It's been a low snow year, which Maletzke explains will likely present its own challenges when it comes time for us to hop on the snowmobiles. He eventually turns from the county road onto a Forest Service road. A small, bullet-ridden sign marks the jurisdictional switch. Near the Forest Service gate sits a squat, ramshackle farmhouse. Outbuildings overflow with the detritus of a working ranch. Cattle graze in a paddock.

It feels as if we're trespassing, but we're not. This is public land, although the ranchers that these cattle belong to most certainly know this dollop of country better than most. Their cattle, which will head out into the hills later this year, serve as a salient visual reminder that the survey work we're doing today will have real world consequences.

Soon the snow on the road becomes too deep to continue in the trucks, so we unload the snowmobiles. A confession: machines make me nervous. Loud ones with no roofs, such as snowmobiles, even more so. I have only the vaguest understanding of how engines work, and the heft and weight of these things, combined with their speed, make them wholly intimidating to me. No doubt sensing this, Maletzke gives me the slowest, steadiest steed. He triple checks that I have put my helmet on correctly.

No one seems too optimistic about the likelihood of success, as it's warm and sunny and hasn't snowed in at least a week. Not ideal wolf tracking conditions.

What comes to mind when you think of a wolf, or wolf pack? Do you picture a line of animals loping along, single file, through deep snow? Or a pack chasing prey, working together to bring it down? Or perhaps pups at the mouth of a den, frolicking in the spring warmth?

None of these images are inaccurate—wolves do behave in such ways—but these sorts of sights are typically only possible in relatively open lands, such as Montana's Lamar Valley. There, tourists gawk from the road using high-powered scopes to pick out the distant wild canines. Otherwise, the only reliable opportunity to catch sight of wolves up close is when they are tagged with radio or GPS collars. Then, biologists take to the air using the marvels of modern technology to find them. Under these unique circumstances, videographers and photographers are able to make crystal clear images of the wolves, images that bring Canis lupus into our homes vividly and intimately. This, in turn, influences how most of us experience wildlife—a high-definition experience, sharp on contrast and fuzzy on context. When I learned to hunt as a twenty-nine-year-old, I was shocked at how small and removed the animals appeared, even when viewed through my binoculars. Naturally, this had much to do with my ineptitude as a hunter, but it also revealed a larger, more important fact: the wild does not submit easily to inspection.

Northeast Washington's wolves, despite living closer to humans than their cousins do in Montana, are harder to find.

Seeing a wolf around here is rare. The woods are thick, the hills steep, and the canines intelligent and wary. "They aren't behind every tree," Maletzke says.

As we motor along the snow-covered roads, he stops often. Leaning to one side or the other—or sometimes getting off and walking a dozen feet—he's looking primarily for paw prints or scat. It doesn't go well. The old snow is a problem, and the bright sun means that any tracks out here are in danger of melting out and losing their form. And in order for Maletzke to estimate the current number of wolves in the Wedge Pack, he will need to find clearly defined tracks.

On top of that, I keep getting stuck.

Maletzke and Samsill race ahead, dodging partially buried trees and navigating steep uneven terrain. At one point, we come to a fallen tree that blocks nearly the entire road. I prepare to stop (we've brought along saws for this very occasion, after all), but Samsill guns it and rides up along the steep road cut like a surfer turning in toward a cresting wave. His snowmobile wobbles, threatening to topple down onto the road and the gnarly looking pine, but by throwing his weight hard to the side, he maintains his balance and guns through the gap. I follow suit, but at the critical moment, as I feel gravity threatening to tip my loud heavy machine over, I panic and turn back downhill, hopelessly tangling myself in the tree. As I rev the engine in hopes of pulling myself free, the rear of the snowmobile sinks deep into the snow.

It takes me a good five minutes, and the help of both men, to extricate myself. I imagine the wolves—or really any animal better-suited to this environment—watching this spectacle while laughing from the cover of the nearby forest. How could such clumsy and ill-adapted creatures manage to so thoroughly dominate the world?

Despite the subpar tracking conditions (and my subpar snowmobiling skills) we do have some luck. Around one o'clock, after several hours of driving, we come to a fork in the road and find a newer set of tracks angling off to the right. Wolves, like humans, prefer traveling by roads as opposed to squeezing through thick brush. We follow the tracks for about a mile, until they diverge into three distinct sets of tracks—a common occurrence, according to Maletzke. Just like that, we've found evidence of three wolves, matching data gathered by other biologists.

Most pack territories look like a wagon wheel. In the center of that wheel is the pack's core range. From this central area they radiate out, patrolling their land and looking for food while keeping other wolves at bay. And so, when biologists survey wolves, they search for roads that either bisect—or nearly bisect—the pack's core area.

"It's not an exact science," Maletzke says of the process. "None of this stuff ever is. But it certainly gives us a good solid trend."

Due to the fact that wolves don't ordinarily stray outside their territory in the winter, Maletzke and other biologists can be relatively confident that the signs they find belong to members of the pack in question.

Wolves mate in the late winter and den in the spring, with pups typically born in late April. Ranchers who lose cattle to wolves often wonder why the state doesn't survey them during the summer, when their numbers are at their peak. Maletzke explains that this is in part due to the difficulty of tracking wolves without snow. But it's also due to the fact that being a wolf is a tough gig, and most of the pups born in the spring won't make it to the first snowfall.

"The population, if you looked at it right after they denned up, you might have a lot of wolves," he says.

Instead, biologists survey packs after a full season, allowing enough time for wolves to disperse or die before a new season begins. The bottom line here is that Maletzke feels confident that the three sets of tracks we've found represent good, solid data. The kind of data that can be plugged into a report, mixed with other data, eventually published and presented to those who make policy.

At the time of the 2019 survey, a minimum of 126 wolves, 27 packs, and 15 breeding pairs lived in the Evergreen State. By 2021, these numbers had risen to a minimum of 206 wolves and 33 packs, according to an annual survey conducted by state and tribal biologists. In the eastern one-third of the state, wolves are protected by endangered species laws specific to Washington; they are federally protected in the western two-thirds. According to the state's wolf recovery plan, wolves can be delisted after fifteen successful breeding pairs have been documented for three consecutive years, or after officials document a minimum of eighteen breeding pairs in one year. Under either scenario, the pairs have to be distributed evenly throughout the state's three wolf management areas. However, the majority of these wolves live in eastern Washington, thus preventing the state from delisting them.

Whenever wolves are delisted from the endangered species list at the state level, wildlife managers hope to have a management plan written and ready to go. Managers anticipate that the plan will be finalized in 2022 or 2023, although no one will be surprised if it takes longer. Disregarding all of these dates and numbers, the important takeaway is that the information we gather today will play a role in determining when Washington State will begin considering wolves as if they were any other species.

Nothing about this process is particularly unusual. All across the world, humans are making ecological, political, and social decisions based, at least in part, on scientifically predicted outcomes. Scientific theory is what has allowed engineers to build giant dams, providing cheap and relatively clean power to millions. Scientific theory is what has brought us modern medicine, electricity, computing, the internet, the internal combustion engine (though perhaps snowmobiles should be considered an aberrant extravagance), and many more modern marvels—marvels that were rare one hundred years ago when the last wolves were killed in Washington.

The shift toward theory is relatively new to the wildlife sciences. Perhaps inspired by the notion of becoming an old-school biologist, many a young wildlife researcher enters the field in hopes of getting out of the office; most are disappointed. Modern ecological studies require far more desk and lab work than they do field work. So much for the fantasy of spending one's days scaling hills, scribbling notes, and sketching flora and fauna.

This sea change in ecological thinking and methodology began early in the 1900s and accelerated in the 1960s. Many of the models and methodologies used today by wildlife agencies across the world were developed by Robert MacArthur, a precocious Princeton biologist who, during his short career (he died at age forty-two of cancer), upended the study of ecology, shifting it from primarily a descriptive discipline toward an experimental one. Alongside the legendary "Ant Man" E. O. Wilson, MacArthur demonstrated that by looking at survival rates, habitats, immigration rates, and other factors, the future fecundity of a species could be predicted.

"His methodological influence on his field of science, though, needs no discounting," writes David Quammen in his book *The Song of the Dodo.* "He changed the way ecologists think. He changed the way they ask and answer questions."

Or, as MacArthur himself stated in the introduction to his final book (written while in the hospital without the aid of notes or reference material): "Not all naturalists want to do science; many take refuge in nature's complexity as a justification to oppose any search for patterns."

He was not interested in these sorts of naturalists, laying out the groundwork instead for those who sought to combine traditional descriptive work with a more theoretical approach. This shift in focus had its detractors. "Ecology for so long glutted with facts but starved for a theoretical structure to organize them is now being provided with a feast of theory it isn't quite ready to digest," wrote one evolutionary biologist in 1975.

Since MacArthur's death, the tension surrounding his methods has persisted, with some questioning whether ecology can truly be a "hard" science in the way of physics. In a crisp eight-page paper published in 2008, Dale Lockwood, a biologist at Colorado State University, argued that many of theoretical ecology's underpinnings are not truly laws.

He specifically targeted the Malthusian growth principal, which essentially posits the idea that populations will grow exponentially if there are unlimited resources for the population and no other constraints. This has been a key presumption used in wolf recovery in Washington and elsewhere, even though, as Lockwood pointed out, no part of this formulation is actually possible. Unlimited resources don't exist. No species lives in a vacuum.

The exponential growth formulation attempts to mimic laws found in physics and other "hard" sciences. And yet, the question of whether these types of ironclad laws even exist in ecology is debated. Lockwood's paper didn't attempt to settle this question so much as highlight the inherent complications in assuming that ecology operates like other sciences:

> Ecology is an important science that is involved with the human condition. It is critical to have society correctly understand that, as a science, ecology can produce results that can inform policy makers and managers to make better decisions. Holding up "laws" that fail to meet the criteria for being a scientific law will not engender a level of confidence in the results of ecological science in general.

In other words, as an earlier critic of the field argued, "It is certainly a function of science to seek order and simplicity, but one old saw of biology is, 'Seek simplicity and distrust it.'"

All of our snowmobile wrestling and wolf sign seeking has been undertaken with one single overarching concept in mind: recovery. Broadly stated, ecological recovery is the effort to bring landscapes back to the state they were in prior to human disturbance—a

Sisyphean task if there ever was one. *How far back* is up for debate, although in the case of wolves it's clear. Up until recently they were absent from most of the western United States.

As for the estimation of when wolves will recover in the Evergreen State, those guidelines were also established by Maletzke. In 2015, he and several other carnivore researchers published a nine-page paper laying out different scenarios for how wolf recovery may unfold. The researchers pulled from a diverse array of material—elk and deer densities (based off hunting harvest statistics), cattle and sheep grazing allotment information, ground cover data from the US Forest Service, among other sources—to build a map of suitable wolf habitat.

Then they applied the information gleaned from wolf recovery in northwest Montana and Idaho to predict where and when packs would form, and for how long the individual wolves would survive. It's a complicated bit of wildlife ecology employing a fair bit of math and modeling in the tradition of MacArthur that is likely unintelligible to the average reader:

$$\text{Log } P_{wolves} = -4.457 + (0.057) \text{ forest cover} + (-0.87) \text{ human density} + (1.351) \text{ elk} + (-1.735) \text{ sheep density}$$

In layperson's terms, this formula helped researchers predict the geographical areas in Washington likely to be recolonized by wolves. In the 2015 paper, Maletzke acknowledged the limits of this sort of modeling: "All population viability analyses are accompanied by uncertainty and only used as guides for management rather than precise, accurate predictions." The study concluded that wolf recovery, as defined by the state, should occur sometime between 2021 and 2023.

Folks on both sides of the wolf issue generally respect Maletzke's work. Still, as we load the dripping snowmobiles back onto the trucks, I'm struck by two thoughts.

First, it's hard not to wonder about the speed at which we've traveled through the landscape. Of course, it makes sense—to cover the kind of ground wolves cover, and still have time for a family and a house, friends, and interests—biologists need to do their research on machines. And for a chronically underfunded, understaffed, and overcommitted agency like WDFW, all efficiencies must be embraced. They're examining landscapes, after all. If they miss a wolf or two in the Wedge Pack it won't matter too much. Agency staffers often have to remind furious ranchers and environmentalists that they're providing a *minimum* count of the wolves.

Second, while theoretical ecology has been a boon to many species and habitats, it has also seen its fair share of flops and failures. These mistakes are generally not the fault of the biologists, but rather the inevitable consequences from the effort to reduce complex systems into manageable parts. The scientific process, of course, is built on a plinth of failure. For researchers, this is an acceptable price of doing business.

But for the people who feel a nearly spiritual connection with wolves, this is cold comfort. Whenever a wolf is killed by the state, the WDFW's assurance that the wolf population continues to grow—that these deaths essentially represent a drop in the bucket—can be rage inducing.

———

Or consider things from the rancher's perspective. Arguments and appeals defending the scientific process don't mean a whole lot when a wolf has gnawed on one of your calves and you're outside at three in the morning trying to keep the wound from going septic.

If a biologist were to approach you after such an experience, and—with no doubt the best intentions—try to explain to you the dynamics of a recovering carnivore population or, heaven forbid, point out that you elected to move into wolf country, you'd be forgiven for asking them to get the hell out of your face.

When we meet in July, Stephen Pozzanghera is sporting long, pandemic-length silver hair and a beard. In consideration of the new epidemiological reality, we chat at a park just a few miles from his office in the small town of Chewelah.

It's sunny, and promises to be a warm day, but the morning is still early enough that remnants of the night's chill remain. Our wooden picnic table is inscribed with the usual graffiti, hearts and vulgarities. Logging trucks downshift, their gears roaring, as they pass through town.

Pozzanghera is living and working here for the summer, having taken an interim position with WDFW aimed at cooling the perennial wolf conflicts in this part of the state. Ordinarily he serves as a regional director, overseeing a wide variety of animal and habitat concerns, but for the next three months his focus has narrowed down to wolves.

His wildlife management origin story will be familiar to anyone who knows a biologist. As a young boy growing up in upstate New York, Pozzanghera was obsessed with animals of all kinds. "I was always outside, always fascinated by wildlife in particular," he tells me. "I did the normal thing, always had critters in my bathtub, you know? One day you were catching butterflies, the next day you were catching snakes."

In the early 1980s, he pursued his undergraduate degree at West Virginia's College of Forestry and Natural Resources, followed in short order by a graduate degree from the University of Tennessee.

As a graduate student, he studied black bear reproduction in the Great Smoky Mountains. "That's where I really started getting interested in large carnivores," he says. His advisor was the famed (a relative term, when it comes to wildlife research) Dr. Michael Pelton.

He went on to work for the state of North Carolina for seven years before taking a job at WDFW in Washington in 1993 working with carnivores. After climbing up the agency career ladder, he took a regional director job on the eastern side of the state in 2010. Pozzanghera's forty-year career has encompassed a sea change in what wildlife management means.

Beginning in the 1950s and 1960s, biologists, ecologists, and state and federal fish and game managers were beginning to understand the full—and mostly terrifying—impact that humans have on the natural world. "We have a responsibility here," Pozzanghera says about the shift in attitude. "Gone are the days where you can simply say that's going to take care of itself."

By documenting the unintended impacts that human innovation could have on our planet, books like Rachel Carson's *Silent Spring* helped drive this evolution in our thinking. Meanwhile, new methods in ecological research—particularly those that combined traditional wildlife management tenets with newly emerging mathematical models—gave fish and game managers a predictive power they'd never had before. When Pozzanghera was in school, these ideas were just beginning to take hold, and he entered into a wildlife management milieu that was still very much focused on maximizing the opportunity for hunters and anglers to fill their tags.

But much has changed over the course of his career. The ideas that were just gaining traction when Pozzanghera was in school are now firmly established doctrines. "I think what we see more and more is a greater shift toward wildlife ecology, landscape ecology," he says. "There has been a shift away from the traditional wildfire management model." That shift has broadened the purview of wildlife managers, who now focus on entire landscapes and ecosystems, not just regional interests. In response, WDFW has broadened its planning horizon.

Every two years, policy and budget wonks at the agency draft a new strategic plan. This cycle is the equivalent of an institutional inhale and exhale, the metronome that underlies decisions and the dates on which budgets thrive—or die. This timeline, which is tied to the state's budget cycle, couldn't possibly be more removed from the needs and concerns of the natural world.

"When you consider climate change, when you consider human population growth in the state of Washington, when you consider all of these factors, are we really doing ourselves and our resources a service by cranking out these biennial strategic plans?" Pozzanghera wonders.

This was the rationale for the decision to shift the agency's planning horizon away from the two-year budgetary cycle and toward a twenty-five-year strategic plan. Pozzanghera's hope is that this will allow the WDFW to better respond and adapt to a changing climate and a growing human population.

As one wildlife employee involved with the strategic planning shift told me, "We have to stop looking right at the end of the hood of the car and start looking at the highway ahead."

This expanded focus is just one of the ways wildlife agencies have changed, moving farther away from the traditional focus on hunting and fishing.

It's a shift that has incited plenty of anger, a shift that has ultimately fanned the flames on most of the recent conflict about wildlife in the state. Pozzanghera tells me that Washington biologists are often accused of being "desk biologists." In rural communities where people interact with wildlife—and wild spaces—as a matter of course, being told what to do by some college-educated biologist who has plugged a bunch of numbers into a computer and returned with a prediction, or worse, a mandate, reeks of naive elitism.

That perception, sometimes valid, often not, is the crux of the issue, Pozzanghera says. "'You're telling me it's based on a model?'" he recalls being told. "'I'm telling you it's based on I've lived here for fifty years.'"

Bridging that gap is part of why Pozzanghera has taken this interim job. As staff biologists and managers do what they do—both in the field and on the computer—Pozzanghera hopes to improve communication with the communities directly impacted by the return of wolves.

Some are skeptical. Scott Nielsen, the president of the Stevens County Cattlemen's Association, called the hiring decision a "slap to ranchers" in a story announcing the move.

"I think they would have been a lot better off with a fresh face," Nielsen said. "I think Steve will be a complete and utter failure at improving relations with ranchers. But, hey, having said that, Steve can prove me wrong."

Unlike Pozzanghera, Daniel Curry did not go to college. He did not study population dynamics. He could not name the four goals of ecological restoration. His work is holistic in nature and terribly difficult to scale, analyze, or oversee. Instead, he's learned what he knows about wolves and livestock protection from his own experiences, first at the wolf sanctuary and then during long days and nights in the saddle.

And so, when he tells me that there is a fundamental disconnect between the environmentalists and the ranchers, I believe that the subtext is this: in reducing the world into laws, in manipulating the land and managing the animals living on it, we've become disconnected and disenfranchised from the natural world.

In big cities, or in areas dominated by monocultures, it is perhaps easy to forget this fact. Nature is subdued, either paved or planted into submission. The situation gets more complicated, however, if you live in a place like Washington State, or much of the American West.

These are places where resurgent carnivore populations are coming into regular contact with humans, and vice versa. Take, for example, the Wedge Pack, smooshed between a highway and a river that looks more like a lake. Wild animals in Washington live nearly on top of humans.

And this causes all kinds of grief, for humans and animals alike. More importantly, it erodes trust in the scientific process. For generations, humans have waged an unending war against nature. We've driven dangerous, bothersome, and not immediately useful animals away. We've poisoned, trapped, and shot them. We've plowed up their homes and paved their pathways.

For some, these wars are not so distant. Rural Americans are less than two generations removed from the time when state fish and game biologists were primarily concerned with deer and elk densities. Not fish and wolf habitats.

Urban Americans, on the other hand, are separated both geographically and temporally from most of these battles. Nowadays, living in their concrete deserts, awash in the excesses of ecological manipulation, they look back fondly on "how things used to be."

Cities tend to hold the political power. And they tend to align with the science and facts. Extinctions are rampant. Climate change threatens the ecological order of things.

And yet, for someone in rural America, the urban policymaking that seeks to address concerns about climate change seems like yet another overreach, yet another step toward their sense of disenfranchisement and disillusion. "How do you think your food comes to you, city dweller?" they ask. "Where do you think your electricity comes from?"

Driving back across the state from the fundraiser at the mansion, this disconnect is on Curry's mind. He imagines what his neighbors would think of the thousand dollar shoes.

"It's a really a disconnected vibe there," he tells me after returning home. "Even more than on the rural side of the state."

Chapter 5

THE CATTLE

Big, flat chunks of steak, cooked to perfection. Grill lines that look as though they've been lifted straight from a glossy advertisement. After all, anything less would be a sin here at the annual banquet for the Cattle Producers of Washington. It's a night meant for socializing, catching up, and talking about the industry's future. Everyone is seated at picnic tables, guarded from the chilly October air by glowing heat lamps.

But where are these steaks from? This turns out to be a surprisingly difficult question to answer. Brian Benson, the owner and operator of the winery at which the banquet is being held, initially informs the gathering that the steak is from Washington. The ranchers sitting around his table laugh. It had better be, after all.

But then Benson, a genial man with a ruddy face and glass of wine glued to his hand, admits that the steaks are, if not from Washington, at least from the tri-state area. The butcher assured him of that.

But even that assurance is dubious, this lack of certainty all too familiar to everyone at the table, underscoring the fragility of the industry. Since 2015, cattle ranchers and meatpackers have not been allowed to label their products as coming from the United States, despite a 2002 law requiring country-of-origin labels to be

---

affixed to all beef, lamb, pork, goat meat, chicken, fruit, vegetable, fish, and a handful of other products.

Known as country-of-origin labeling, it's viewed—at least by US ranchers—as a key competitive advantage. Many consumers are willing to pay more if they know that the meat they're buying comes from the United States.

However, in 2009 Mexico and Canada filed a complaint with the World Trade Organization alleging that the US law discriminated against their products, and in 2012 the WTO ruled in their favor. Meanwhile, several major US meatpackers also objected to the country-of-origin labeling mandate, citing the increased costs associated with separating foreign-raised meat from American-raised meat. In response, the Obama administration repealed mandatory country-of-origin labeling requirements for beef and pork. Retailers and producers are still permitted to label the origins of their beef, but most ranchers sell their cattle to a series of middlemen, exiting the supply chain long before the animals make it to either the slaughterhouse or the packing house, which means that the ranchers effectively have no control over how their meat is labeled.

In 2021, a bipartisan group of senators from Montana, South Dakota, and New Jersey introduced a bill called "The American Beef Labeling Act of 2021" that would reinstate country-of-origin labeling for beef products. As of May 2022 the bill remained in the Committee on Agriculture, Nutrition, and Forestry.

All of which is to say that the steaks we tuck into at the cattle producers' dinner could easily be from Uruguay. Or from Argentina, or Australia, or from any of the twelve other countries that produce most of the world's beef. Ultimately, the ranchers who raise cattle in

Washington and other Western states don't really know where their animals are going. And even the savviest of consumers, dedicated to buying locally grown meat, cannot be confident in actually knowing where their steak comes from. (There are some complicated ways around this, but for most ranchers, who are already working with tight profit margins, these workarounds aren't possible.)

And that, argues Bill Bullard, the executive director of R-CALF USA, a rancher advocacy organization, has everything to do with wolves. "Your cattle industry has never been as precarious as it is today," he thunders at the assembled ranchers. "You are a skeleton of what you were just years ago."

There are murmurs of agreement. What's more, Bullard continues, this skeletal apparition means that non-ranchers don't know, or don't care about ranching. And that, in turn, means that ranchers are increasingly isolated and marginalized.

Case number one, Bullard posits, are wolves. If small, independent ranchers were to have more economic clout, they would have more social clout, and in turn, they would be better positioned to advocate for their interests. Instead, they're getting beaten by the environmentalists. By the state wildlife agencies. Steamrolled by more relevant and powerful lobbies. Bullard's on a roll now. Pacing back and forth in front of his audience, his thick mustache the perfect prop for this talk. "And what," Bullard asks, "is the underlying cause of all this?

"Broken markets."

When Scott Nielsen, the president for the Cattle Producers of Washington, addresses the assembled ranchers, he briefs them on the CPOW livestock monitoring program.

In the spring of 2020, CPOW got a grant from the Washington Department of Agriculture to hire range riders. The program, which was already being employed by the state wildlife agency and a regional collaborative funded by a conservation group, aims to provide human presence in wolf country. During its brief existence, range riding has had its fair share of controversies, not least of which was when the state accused two range riders of shirking their duties, shopping in a nearby city while collecting paychecks from the state. While those riders were allegedly gallivanting around, wolves were killing cattle.

In short, there is an ongoing trust issue.

What's more, range riders face the unenviable task of wandering around a rancher's operation. For ranchers, this is the rough equivalent of allowing someone to poke around your backyard day and night. This is the sort of relationship that requires trust. And, as Nielsen puts it, "I wasn't interested in WDFW or Conservation Northwest coming out to my operation. I wanted them to be accountable to me."

For these reasons, CPOW started a livestock monitoring program aimed at identifying "all the carnage that's happening out there." Conservation and environmental groups are deeply skeptical of the endeavor, believing—with some justification—that CPOW livestock monitors are simply checking boxes. In fact, at the ranchers' annual meeting, the assumption that these monitors are sympathetic to ranchers was no secret, and served as a selling point for the program.

"It was the difference between life and death for us," rancher Ted Wishon says about the program.

Long before he was born, border politics landed Wishon's family in Washington, when his grandparents tried to drive a herd of horses to Alaska but were turned back at the Canadian border. His father ran a sawmill and then Wishon started ranching in the 1980s. There have been plenty of ups and downs, but in recent years the downs seem to come more frequently. A big man with a commanding voice (a common trait among those who are accustomed to directing five-hundred-pound animals), Wishon places some of the blame for these struggles on the legislation regarding country-of-origin labeling. But he also believes that wolves have bitten into his bottom line. Still, he has declined to use state-hired range riders or nonprofit riders, not wanting to take their money and have the "enviros holding our purse strings."

As far as he is concerned, the CPOW program has eased these worries and has even convinced him of some of range riding's core tenets. And, somewhat cynically, the CPOW program allows ranchers like Wishon to say they've tried to do everything they possibly can to prevent wolves from killing livestock, which in turn forces the state to kill problematic wolves.

"Human presence sounds pretty corny, but it has some impact, I think," he tells the group. "Other range riders, if they see something you're not proud of, like a cow in a ditch or something, they will report you right away. CPOW riders don't. They're on the rancher's side."

Which is exactly why many conservation and environmental groups have criticized the CPOW program. There isn't enough oversight, they argue. In their view, "livestock monitors" are covering too large of a geographic area to be effective, and that at the end of the day, these riders don't care about wolves.

"We need to ensure that proactive range riding is happening up there," Zoë Hanley, the Northwest representative for Defenders of Wildlife, will tell me later. "If there are people up there who are really interested in true conflict mitigation, these community-based initiatives are the future. We just need to make sure that accountability is in place."

For his part, Daniel Curry dismisses the entire endeavor. "The whole thing there is a rigged carnival game," he says of the CPOW riders.

When dinner ends, the group of ranchers files back into the meeting room to listen to a "cowboy comic." I decide to skip this bit of entertainment and head back to my car. As I'm walking out the door, Scott Nielsen pulls me aside. He tells me that he's annoyed with some of my recent wolf coverage. "What coverage in particular?" I ask.

"You keep quoting state employees," he says. "They're all liars."

I leave the meeting feeling bewildered and frustrated. Driving home along winding rural roads, I can't stop thinking about how one of the issues that environmentalists are most concerned about—livestock monitors representing the interests of the ranchers, not the wolves—had been proudly advertised as fact.

How can any sort of reasonable compromise ever be made with two such divergent belief systems? My lack of optimism is only compounded by the sense that, despite all his efforts to the contrary, Daniel Curry's inroads to the ranching community seem to be crumbling, soured by the politics of wolves and humans.

THE RETURN OF WOLVES

Exhibit A: his alliance with Jake Nelson.

Two years earlier, late in October, Curry started his workday at 10:00 p.m. As he saddled his horse, the moonlight illuminated an abandoned homestead, now surrounded by aged paddocks and cheatgrass. A century ago, the Dulin family lived and ranched here. Now the Grumbachs, one of the area's largest ranching families, owned the four hundred acres.

Once he was ready, Curry headed up the hill, the warm plume of his breath visible in the chilled air as he navigated through nighttime obstacles. He'd brought along a saw to cut trails, a shotgun with non-lethal rounds to scare off wolves, and a GPS device to track his work and mark where he saw wolf sign.

Although the Grumbachs owned the land, the cows themselves were owned by Jake Nelson, a rancher who was twenty-seven years old at the time. Nelson himself comes from a family of ranchers based in Chesaw, Washington, a former gold-rush town located northwest of Republic, the county seat of Ferry County.

At the time, Nelson was engaged to Doug Grumbach's daughter Amanda, and the young couple had decided to take a stab at being twenty-first century ranchers. Ranching was what Nelson had wanted to do forever.

As in all forms of agriculture, the power in ranching has consolidated, with larger outfits taking over market share and young ranchers like Jake and Amanda becoming unicorn-level rare. And so other ranchers celebrated the young couple's efforts, even if Nelson's views, not to mention his willingness to work with Daniel Curry, were a tad nontraditional.

During the 2018 grazing season, the state confirmed that Nelson lost three of his cattle to wolves. But he believed—and some research supported this—that for every dead bovine there were three more that weren't found. Before meeting Curry, Nelson held a dim view of wolves: *Heck no. Get out. This is my livelihood at risk.*

But Curry's persistence, dedication, and competence in the woods slowly won Nelson over, and for several years, the two men worked closely to try to keep wolves away from cattle. They spent long hours together, getting up before sunrise to wrangle cattle and going to bed well after dark. Curry learned the lay of the land and worked with Nelson's cattle, all while keeping tabs on the resident wolf pack. Nelson despised the Washington Department of Fish and Wildlife and viewed other ecologically oriented organizations with only slightly more respect.

"I would have to say that I think the wolf is a pretty cool animal," Nelson told me once. "The wolf isn't the problem. It's the management that's the problem."

This view of wolves, as vanilla as it may sound, is actually a hard-won admission of acceptance from a rancher. In fact, many of Nelson's fellow ranchers would view such a statement with skepticism, particularly when it's been shared with a reporter.

And Nelson is hardly a wolf-loving hippie. A foul-mouthed and hard-drinking man, in August 2018 he learned from GPS collar data provided by the state that a wolf had been sniffing around his cattle. Nelson went to scare it away, only to see some pups with the adult male wolf. According to a state news release about the incident, the male wolf growled and started to approach—typical behavior of wolves with pups—and Nelson shot it, breaking the wolf's leg. About a week later, the state killed the injured wolf.

This incident pained Curry deeply. That wolf was from a pack he knew, and he didn't believe it needed to be killed. But he stuck with Nelson, continuing to help him, singing his praises, telling any-one who'd listen that "Jake is trying." This was the rancher with whom we'd howled at the moon. A man Curry considered a friend, even if they disagreed vigorously about many things.

But by 2020, things had soured between the two men to the point where Nelson no longer answered Curry's phone calls. Some of this distance may be attributable to the fact that Nelson and Amanda Grumbach were in the process of divorcing; perhaps Nelson had given up on the idea of coexisting with wolves and didn't want to tell Curry directly. Either way, Curry was assigned to a different allotment of land that spring. Some of the blame for their relation-ship's deterioration can and should be hung on the shoulders of the reactionary and conservative ranchers and cattlemen associations who unfairly blame wolves for all their problems. Daniel Curry has some inkling of what happened, although it's hard for him to be cer-tain because Nelson doesn't pick up his phone calls. Perhaps most tellingly, Nelson is now working with CPOW, hiring fellow ranchers to guard against wolves.

Underlying all of this is a generalized distrust in the powers that be. That's why ranchers feel compelled to form their own livestock monitoring operations and why Jake Nelson is disdainful of state agencies, distrustful of the media, and furious at urban environ-mentalists. And why shouldn't he be? After all, blue-collar workers across this nation—whether they're ranchers, welders, or miners—have been sold down the river too many times to count.

When former President Donald Trump intoned "Make America Great Again," northeast Washington listened. He was speaking to their anxieties, no matter how callous and dissentious his real intent may have been. In an echo of what took place with the matchmaking industry in Diamond City, rural Americans have seen their ways of life vanish.

I grew up around the rubble of such vanishment.

These days, Coeur d'Alene, Idaho is known as a resort town tucked into the northern end of a massive lake. During the summer, any number of languages can be heard in the lobby of the Coeur d'Alene Resort while vacationing jet setters check in and out. Justin Bieber, Oprah Winfrey, Shaquille O'Neal and other celebrities own cabins—mansions, really—on "the Gem Lake," occasionally even playing beer pong and mixing it up at local bars, as Shaq did on New Year's Eve in 2016.

Drive forty minutes east on I-90 and the glitz and glam are nowhere to be found. This is the Silver Valley, once one of the most productive mining regions in the world, churning out billions of dollars worth of lead, silver, and zinc over the course of a century. This success came at a grave cost; a series of disasters befell the valley, including a lead-poisoning epidemic of children in the 1970s.

Starting in the 1980s, a newly unified global trade network made mining in other countries more profitable for multinational companies focused on maximizing profit. Meanwhile, the newly environmentally conscious US public started to worry about what the long-term cost might be of resource extraction on such a grand scale, and so the companies that had vacuumed up wealth from the mountains of north Idaho moved on to richer, less bureaucratically complicated lodes.

This left behind a community that lacked the infrastructure to make a living outside of the mining industry. Many people lost their jobs. It was around this time that my hometown of Coeur d'Alene pivoted to tourism, which has paid off handsomely in some ways. However, the average wages for workers in the service industry pale in comparison to those who work in mining or logging. The end result is a valley dotted with ghost towns. The best-case scenarios are the towns endlessly reliving past opulence in the forms of historic tours through once-deadly mines, brothels, and theaters.

As a thoroughly mediocre high school basketball player, I remember traveling to one such shell of a town. Their gym seemed mammoth and opulent to me—a gym that could easily sit seven hundred people. And yet the entire school district, kindergarten through twelfth grade, had only one hundred students.

And so, when Bill Bullard, the fast-talking cowboy turned politician turned advocate, booms on about broken markets, he's speaking not just about ranching, but also about this great contraction.

"We are losing cattle producers," warns Bullard. "We are losing feedlots. We will soon reach the point of no return."

Consider the case of "Cattlegate," or if you're a federal investigator, the "ghost cattle" scam. Two hundred thousand made-up heifers. A massive fraud rocking eastern Washington's arid ranching communities, leading to criminal charges and bankruptcy. The Church of Jesus Christ of Latter-day Saints and a Bill Gates-owned company duking it out at the auction block, each willing to spend more than $200 million to buy 22,500 acres of ranch land and its associated water rights.

These were just some of the headlines in 2021 when Cody Easterday of Mesa, Washington, pleaded guilty to defrauding Tyson

Foods and another unnamed company of more than $244 million. According to court documents, he did so by billing for the care of these nonexistent animals. Then he used this ill-begotten money to pay off the debts he'd incurred from gambling on the future price of beef.

Why did Easterday do this? Speculation is rampant, in part because he's declined all media requests at the time of this writing. In addition to his shortcomings as a businessman, and his ethically unsound decision-making, there are rumors of a gambling problem. But there is also a deeper, systemic issue at play in his story—an issue of markets.

For decades, ranchers (and American farmers in general) have watched as their already anemic profits have been further gralloched by larger and larger corporate outfits. They've witnessed free trade agreements being enacted that end up pitting American farms—and all the attendant labor standards—against other countries with (ahem) looser regulations. "You can be the best farmer in the world and still go broke," a wizened apple orchardist once told me.

The mechanism by which the ranching industry has been bled is particularly fiendish. For decades, ranchers competed in a cash market. That is, they negotiated the price of their cattle with the meatpackers before the sale.

However, with international trade deals flooding US meatpackers with cheap beef, domestic producers lost important leverage. And, due to the fact that the timing of the slaughter is all important—meaning a rancher gets less money if the animal is too big, or too small—the meatpackers have the upper hand. As the market was flooded with cheaper beef from other countries, America's meatpacking industry consolidated. Currently four corporations control 73 percent of the beef in the United States.

To maximize their profit—and because they have the upper hand in this arrangement—the packers started negotiating contracts ahead of time, contracts that ensure the producers will sell their cattle in a timely fashion. The upshot of all of this is that meatpackers will loan cattle producers money in exchange for an agreement to buy their future cattle at premium market prices.

The catch? The producers don't know how much the packers will pay for the cattle. This is known as "formula contracting," and meatpackers argue that it provides ranchers with steady access to markets, income, and bank loans.

"Cattle producers are willing to give up a known price in order to have a timely access to the market," Bullard says. However, because the US beef industry is so centralized, there is little to no competition on the open market. Many argue that this in turn leads to depressed prices.

A 2021 story in *High Country News* contains some eye-opening details:

Ranchers have long complained about lowball prices from these companies. Nationwide, data from the United States Department of Agriculture shows they have reason to. Profits for ranchers have trended slimmer almost every year since the late 1980s, when those prices were first tracked. By 2020, the same year the Easterday empire began to crumble, a rancher's share of the value of boxed beef shipped to retailers was 37.3%, down nearly 27% since 2015, when it was

51.5%. This while the consumer price of beef soared higher than ever.

This is the financial reality plaguing Cody Easterday and ranchers in general. In hopes of alleviating that burden, Easterday started betting on the future price of cattle on the Chicago Mercantile Exchange. Like a stock bet, he bought futures in beef, hoping that doing so would offset any losses sustained in the direct sale.

But, like any speculative endeavor, there were ups and downs, more downs in fact. By the time he pleaded guilty to defrauding Tyson Foods, he had lost more than $200 million in the futures market.

After his guilty plea, the bidding war on his land started. In 2022, the Church of Latter-day Saints' agricultural holding company beat out Gates' 100C LLC, cementing the Mormon Church as one of the largest commercial agricultural landowners in the West. This story of consolidation has been mirrored across the country; in 1987 more than half of all US cropland was operated by midsize farms holding between 100 and 999 acres of cropland, while only 15 percent was operated by large farms with at least 2,000 acres, according to a US Department of Agriculture report in 2018. Over the next twenty-five years, those numbers shifted dramatically. By 2012, farms with midsize holdings only held 36 percent of cropland, the same share as that held by large farms.

Ranchers like Jake Nelson are aware of these trends. Even if they can't rattle off the exact numbers (although many can), they certainly feel the change in their bones, having watched so many neighboring businesses fold. They've seen ranches sold to bigger outfits. They've seen ranches snapped up by developers and parceled into forty-acre plots designed for the sensual pleasures of city-weary folk. They've watched as the price of beef goes up and up, while the amount they're paid in return remains the same or even falls.

The "powers that be" have continuously assured them that this is all in their best interests. Economists go on and on about how free trade is ultimately a win for the American worker. Politicians fetishize blue collar iconography—big trucks, elbow grease, yada yada—all while the actual people who used to do those jobs now find themselves cleaning millionaires' second homes, turning down sheets in swanky hotels, or selling the land their family has worked for one hundred years and seeing it subdivided into new homes.

So of course they listened when a politician spoke directly—and plainly—to their plight. And of course many of them were let down when Trump revealed himself to be just another grifter.

Wolves have become the unintended victims of this legacy of distrust. The assurances of biologists and state agencies—who attempted to convince ranchers, farmers, and others who worked in agriculture that wolves would be good for the local ecology, that wolves wouldn't kill (too many) cattle, and that wolves almost never attack humans—sounded familiar to the people of eastern Washington. They'd heard this kind of pitch before. Their distrust was exacerbated by the knowledge that those who were happiest about the return of wolves were urbanites—people who have done well financially even as the fate of the average rural American worker has spiraled.

Daniel Curry knows this history. He knows these stories. And he had hoped that by leading by example, by moving into wolf country, by working alongside the ranchers, and by raising his own livestock (sheep, in his case), he and the ranchers would come to an understanding, and he would be able to affect change.

But there is a line he won't cross. He won't sign off on the death of wolves when there are other options. That's because the biologic science of wolves is clear: they are a necessary part of the landscape, and coexistence is possible.

Chapter 6

PREDATORS AND PREY

The deer in question is turning pungent in the summer heat, and is unmistakably dead. But, as with any murder investigation, establishing the fact of death is the easy part. Instead, Taylor Ganz, a PhD student at the University of Washington, is interested in the how, the when, the why, and most importantly, the who of this crime scene.

"I'm always hesitant to even state my first impression, because I don't want to ultimately bias the decisions," she says, moving slowly and examining every inch of the forest around the dead cervid.

She learned of the animal's death via text early this morning. The deceased doe was one of the one hundred or so deer and elk that Ganz is monitoring in eastern Washington using GPS collars.

Ganz is trying to understand the tangled web of connection between predators, prey, and, increasingly, humans. Her research supports one part of a statewide study that began in 2016 when the University of Washington embarked on an ambitious project in conjunction with the Washington Department of Fish and Wildlife, in which they hoped to document the various connections between humans, predators, and prey and how those relationships are changing with the return of wolves.

Titled the Washington Predator-Prey Project, the large survey seeks to inform management of the "myriad species that comprise these systems."

Earlier today I met Ganz and her research assistant, Clara Hoffman, at a small home in Chewelah located next to a thrift shop and across the way from a public school. Ganz has spent several summers and some winters living in this house alongside as many as ten other researchers. This is her field team, and they spend their days poking, dissecting, and generally investigating dead animals, primarily deer and elk.

Piling into a university-owned truck, we drove out of town and onto a scraggly patch of land owned by a local timber company. Having been logged in the past few years, the open fields are tricolored, with the pure white of blooming yarrow, the mustardy yellow of St. John's wort, and the green of pines and grasses.

This sort of landscape, a patchwork of timberland and private holdings, is part of what makes Ganz's work unique and important. Nowadays, wild animals live in what Ganz calls "complex" landscapes, ones defined by human interests. Practically speaking, this means that wild animals can no longer be solely relegated to protected natural areas, parks, or tracts of wilderness. "There are only so many parks out there," she says. "To truly recover them, you are going to need to be able to manage these animals in human-occupied landscapes."

Ganz parked the truck, making sure it was well off the road. She's worked in northeast Washington for four years and in that time has slowly made inroads with the locals—inroads she's careful to protect. "I think her carcass is in those trees," she said, gesturing toward a thicket. "Hopefully."

Although modern tracking technology makes it easy to find these dead deer and elk on a map, getting to them can be a trickier proposition. Ganz and her crew regularly hike miles through steep and brushy terrain. It's hard, smelly work.

But Ganz has the background for it. A Los Angeles native, she went to college in Portland, where she studied physics. After graduating, she worked for the National Outdoor Leadership School in Landers, Wyoming. There she became a proficient backcountry traveler, adept at navigating harsh conditions.

While working for NOLS, Ganz assisted with some research for the US Forest Service, a stint that ultimately changed the direction of her life. In particular, she designed and conducted a study monitoring snowmelt and air pollution in Wyoming's Wind River Range. Her work, which isn't published, involved making two trips into some of the region's most remote alpine lakes. From there, she earned a master's in environmental science from the Yale School of the Environment.

Within a minute or two of picking our way through the underbrush—a journey made treacherous by limbs and stumps leftover from the logging—Hoffman, the research assistant, spots the dead animal. The doe's head is just barely visible, buried as she is under a pile of pine needles and duff.

Storing their prey for later is classic cougar behavior. But Ganz is a scientist, and so she hesitates to make any pronouncements until she's fully examined the kill site. "Right now, we're looking at the big picture," she tells me. "The first thing I noticed was that it was well cached. That's typical of felids." She investigates outward from the doe's corpse, looking for drag marks, prints, or fur. She stops and picks up some tufts of grayish fur, then takes a strand and squeezes the ends together, watching the fine hair arc, bending but not folding, a telltale sign that this is predator hair. Ungulates have hollow hairs that keep them warm. When folded, they bend into sharp right angles.

All the evidence is pointing toward a cat, likely a cougar. It's the right kind of habitat—treed. And the carcass isn't strewn about. Canids tend to be messy killers. Felids are more surgical in their approach; consider the difference between how a house cat eats its dinner and how a dog goes about it. Things are not so different in the wild.

After surveying the surrounding area, we move to the carcass itself. The doe is covered neatly under a pile of pine needles, with only her head and leg visible. After uncovering the deer, Ganz and Hoffman deconstruct the animal, starting with her skin. As they do, they note bite marks, broken bones, bleeding, and the internal state of the animal's organs. Blood is pooling where the doe's left chest cavity has been ripped open. The deer's rumen, a large sack that stores food prior to digestion, has been carefully extricated, another clue pointing toward the delicate violence of big cats. Ganz examines the doe's organs, before taking a look at her bone marrow and fat—all to get a picture of how healthy the animal was before she died.

This makes for a gory, smelly autopsy. Minutes after the researchers uncover the carcass, flies and wasps are swarming. All told, the process takes an hour. Once the animal is skinned, Ganz and Hoffman look for signs of hemorrhaging, observing and measuring the width of bite marks. Only after finding two clear bites, and measuring the distance between them, does Ganz feel comfortable in confirming this as a cougar kill.

The granular, painstaking attention to detail is vital. Only through careful examination can the researchers rule out the possibility that the animal might have died from a disease, before being scavenged by another animal. Or the possibility that the doe was already sick or malnourished, and therefore more easily killed.

One of the primary areas of focus of the predator-prey study is on direct mortality; in other words, the researchers are gathering information about how deer and elk are dying, what's killing them, where and when. Additionally, the researchers hope to understand how these various predators are affecting one another, and how this in turn is changing the behavior of their prey.

"We want to understand how wolves are impacting deer and elk populations," Ganz says. "We know wolves eat deer, of course, but it's more complicated than that, because wolves impact other carnivores who also eat deer or impact deer. Wolves change deer behavior. And so, by going around and classifying these mortalities, and understanding how they died, we can get a better sense of what the primary causes of mortality are for this population."

Ganz notes that in four years she's only documented one confirmed wolf kill, and another possible one, despite putting out several hundred collars. This doesn't mean that wolves aren't killing deer and elk—they most certainly are. Just not to the extent that many believe.

"There is a lot more cougar predation I see out here than wolf," she says. "There are a lot of vehicle collisions."

All of which flies in the face of what the folks on the ground— hunters and farmers—believe is going on. They maintain that the return of wolves has decimated prey populations. Ganz is aware of what the locals think; they share their thoughts with her all the time. Herein lies one of the underlying tensions for any modern wildlife biologist: she's looking at the big picture—at populations and large ecosystems—while many of these folks are focused only on what they have seen in the back forty.

When he first came to the University of Washington in 2008, Aaron Wirsing was a shark expert. While doing his graduate and doctoral work in Florida, he often traveled to the islands of French Polynesia, one of the world's largest shark sanctuaries, to study how the finned predators impacted their prey (in addition to killing them).

Wirsing planned on continuing this line of study when he landed a tenure-track job at UW. But 2008 happened to be the same year that a wolf pack first returned to Washington State, and he quickly realized that he had a unique chance to study, from the beginning, the return of an apex predator to a landscape full of humans.

Ecology is a relatively new scientific discipline, and most of the research that's been done so far on large carnivores, like wolves, has focused on relatively wild places. Great insights on how animals behave in intact ecosystems have been gleaned from places like Yellowstone and the interior of Canada or the Arctic. However, these studies have focused on predators living far from people.

Washington is different. In 2019, a wolf was shot and killed just forty miles from Spokane, the state's second-largest city. Unconfirmed sightings have been documented even closer than that. Furthermore, wolves aren't the only large predator found in Washington—bears, cougars, and lynx also roam the Evergreen State. "Washington has become such an exciting area for large predator research," Wirsing says.

Since 2008, he and his colleagues and their students have looked at everything from how wolves impact the movement behavior of deer to how humans impact the behavior of wolves. Their research is important for its inherent scientific value. But it's also helping lay the foundation for a new way of understanding and interacting with the natural world.

"We have to better understand the roles that predators play," he says, "and also the ways humans mediate those roles."

Wirsing tells me that the discipline of ecology only really "ramped up" in the 1950s after "a time of really intense marine and terrestrial harvest." Due to this late start, he believes that our general understanding of how animals live and behave, particularly predators, is, in his words, "stunted," relying on faulty (or just plain wrong) baseline assumptions. "Much of what we know about predators is based on science from after we suppressed predator numbers globally," he says.

Washington is the perfect laboratory for him, then, to study the return of predators in the Anthropocene. Wildlife managers and scientists from around the world have visited. He's had students do research in Africa, where the species are different but the problems are similar. "What we learn in Washington is actually applicable to so much more of the globe," he says.

On a chilly, fog-blanketed morning while camping with Daniel Curry, I end up seeing that global interest up close. Crawling from my tent, I emerge only to find that Curry has already left—patrolling for wolves, I presume—and that his horse Griph is gone, too.

I make myself some coffee and decide to take a walk, shaking off the chill from my legs. The November fog is just starting to break, the sun peeking through, when a truck crests a nearby hill and turns into the driveway leading to the old barn we've slept near.

Knowing the opinions many locals hold about wolves, and journalists, I watch the truck warily and stay out of sight. Perhaps I am overly paranoid. The truck parks, and I am relieved to see Jay Shepherd, Curry's boss, stepping down. A former biologist who managed wolves for the state, Shepherd now runs a nonprofit focusing on cattle-wolf coexistence.

As I'm making way over, another man gets out of the truck, someone I don't recognize. When I reach the truck, Shepherd introduces me to his companion, a Nepali wildlife biologist named Lanzhoujia Ihadrukgyal whose work focuses on the coexistence between humans and snow leopards.

He's visiting Washington to learn from a region that's leading the way in the field, with all the attendant successes and failures of trailblazing. We end up talking for twenty minutes, and the similarities are enlightening. In Nepal, scientists from the city must convince villagers that snow leopards don't necessarily have to be killed, and villagers live with the consequences of having large predators in their backyard. The story would sound familiar to any eastern Washingtonian: an ecosystem that's adapted to not having a key apex carnivore readjusts to the old normal. Documenting the particularities of this sort of shift—from a near predator-free landscape to a predated one—is yet another distinction that makes the Evergreen State unique.

Like many other wildlife ecologists, Laura Prugh started studying wolves in Alaska's Denali National Park. In particular, she investigated how wolves (and other large predators) impacted smaller predators called mesocarnivores.

After finishing her postdoctoral work, she worked at the University of Alaska Fairbanks for nearly four years before heading south to the University of Washington. Since 2016, she's worked on the Washington Predator-Prey Study.

Wolves have famously and not quite accurately gained the reputation for being a great ecosystem balancer—a panacea against all sorts of excess. Most notably, biologists and ecologists noticed that heavily eroded streams in Yellowstone started to heal after wolves were reintroduced in 1995. What was going on? One theory, which went viral in a YouTube video entitled "How Wolves Change Rivers," argued that wolves reduced the elk population, which in turn gave damaged streams a much-needed respite from being eroded by constant foraging from voracious elk.

This simple and easy to understand hypothesis has been attractive to wolf advocates and others looking for basic explanations to complicated issues. Unfortunately, it's not completely accurate. More nuanced research indicates that the return of beavers may have played a larger role. Still, the resurgence of beavers may be partially due to the presence of wolves, with some research showing that when wolves returned to Yellowstone National Park, they culled an overgrown elk population (from 19,000 elk in 1995 to just 6,200 in 2021), which in turn allowed aspen and willow groves (favored foods for beavers) to regrow.

However, even that slightly more nuanced hypothesis has been challenged recently, with researchers from Utah State University arguing that the initial aspen study overestimated the rate of aspen regrowth.

This is not to say that wolves haven't had a beneficial impact on Yellowstone and other ecosystems to which they've returned. As one researcher involved in the aspen back-and-forth told a Montana journalist, "Ecological research is messy and often requires long-term data and rigorous analyses to get the story right."

Wolf research tends to get weaponized, whether by advocates or detractors. Studies that show wolves to be great ecosystem balancers are trotted out by the advocates. On the other hand, ranchers and hunters point to other research indicating that wolves have the potential to decimate prey populations when their growth is left unchecked. Meanwhile, serious biologists continue to do the slow, messy, and sometimes nonlinear work of field research. The goal of the Washington Predator-Prey Project is to parse this mess, and this kind of effort doesn't typically lend itself to simple talking points.

Some of Prugh's preliminary research indicates that the presence of wolves and other large carnivores has the impact of decreasing the number of smaller carnivores. Simply put, the smaller animals try and scavenge wolf kills and in turn are killed by wolves. For famously robust and over-populated species like coyotes, the return of wolves, bears, and cougars has provided a natural limit to their growth. This, in turn, may have positive downstream effects on other species, such as ground-nesting birds.

"I've always been interested in how species interact with each other," Prugh says. "Predator-prey interactions. How a change in one part of an ecosystem can affect other parts through these indirect pathways."

Smaller carnivores are, for the most part, equal-opportunity eaters. "They tend to have widespread impacts on ecosystems," explains Prugh. Sage grouse, the endangered ground-nesting birds that once flourished in Washington's arid plains, are now competing with large coyote populations. Human-led efforts to kill coyotes are mostly ineffective.

"If scavenging increases the risk of mortality of smaller carnivores, that might explain why it appears to be very hard for humans to replace the role of large carnivores in a landscape," Prugh stated in a 2020 press release that accompanied the publication of a paper touting the preliminary results of the predator-prey project. "This link between scavenging and mortality might be one of the mechanisms that make large carnivores so effective in controlling smaller carnivores."

Prugh's findings could potentially have wildlife management implications, and might point toward one option for controlling the populations of smaller carnivores—wolves. "If we want to keep [mesocarnivores] at lower levels, that diversity becomes pretty important," she says. Regardless, she suspects that wolves aren't having as big of an impact in Washington as they've had elsewhere. "In Yellowstone," she says, "they are able to reach densities as set by their prey, rather than by human activity and habitat changes."

The Washington Predator-Prey study is also investigating how cougars respond to wolves, and vice versa. Lauren Satterfield, a University of Washington PhD candidate studying with Wirsing, captains that research.

Satterfield's interest in cougars began in India and Nepal. During her time as an undergraduate majoring in mathematics in Massachusetts, she had the opportunity to spend a semester on the Indian subcontinent. While studying abroad, she did basic fieldwork surveying the different ways livestock owners tried to keep snow leopards from eating their animals. Although the species and societies are different, predator-human coexistence issues in the Himalayas look remarkably similar to those in Washington. After returning to the United States, Satterfield continued working with big cats, first in Montana and Wyoming, and more recently in Washington.

Cougars are not rare in Washington. This is an intentionally vague statement. Although cougar sightings and cougar attacks on livestock have increased dramatically over the past years (primarily in northeast Washington), researchers caution that these reports don't necessarily mean the cougar population has grown.

For anyone who has lost animals to cougars, or been stalked by them, or who believes that cougars are killing all the deer and elk, this assertion will feel counterintuitive, even infuriating. *How can you say there aren't more cougars?*

Here is what some biologists say: Cougar populations have remained steady. On average, a male cougar's range is between fifty and one hundred fifty square miles. They're territorial animals; they don't share space well. In scientific terms, they reach their "carrying capacity" quickly.

What's changed, according to these biologists, is that as we humans have expanded our footprint, we've encroached onto traditional cougar territory, which means more sightings and encounters. Add to that a recent spate of high-profile attacks around the West, including one death in Washington, and this has resulted in an increased awareness of cougar activity, which in turn has led to more reported sightings.

This is due to a well-documented cognitive experience called the Baader-Meinhof phenomenon, also known as the "frequency illusion." Essentially, as soon as you learn about or experience something new, you're more likely to notice that same thing in the future.

On top of that, the proliferation of game cameras has revealed a whole new world of animal movement to us, a world that had been previously invisible.

Other biologists and wildlife managers disagree with this theory, arguing that these increased sightings are due to the fact that cougar populations actually *are* growing. They believe that cougars are responding to reduced human hunting while simultaneously repopulating old habitats. Hound hunting was outlawed in Washington in 1996, and other methods of hunting cougars are more a matter of luck than skill.

All of this ultimately underscores the point that little is known about these predators. Nevertheless, everyone does agree on two things—that there are more cougars than wolves, and that no one really knows how the two species will respond to each other after decades of separation.

This is where Satterfield's interests and expertise come in. The questions she's posed are diverse: Will cougars move to higher elevations and begin stalking different kinds of prey? Will cougars have to work harder to get a meal, possibly reducing their numbers?

These questions—and many others—haven't been answered yet. The predator-prey study has concluded fieldwork, and some of the first papers have been submitted to journals.

No doubt these findings will be followed by many peer-reviewed articles in prestigious journals. But the results of the predator-prey study will also be important to the legions of people who don't eagerly watch for the latest ecological research.

The bottom line is that this sort of research highlights the importance of addressing the trickiest and most difficult dimension of wildlife ecology and management, the part that Daniel Curry has been struggling with for years.

Humans.

In early September, PhD student Taylor Ganz returns to the spot where the doe was ambushed and killed by a cougar one month earlier.

The heat of the summer has had its way with the vegetation and the land is dry. Each step brings with it the crackling premonition of a wave of fires about to sweep the West. We dig through the underbrush, finding the few remains of the hapless doe. A spinal column, grown furry as strands of bone marrow unravel under the wind and sun. A few patches of fur. A hoof.

Today's mission, however, is not animal-related. Instead, Ganz wants to document the types of plants and trees found in this patch of woods. The area has been logged repeatedly, and is currently owned by a timber company. The stands of young-looking trees will likely be logged yet again.

———

"What are the food resources available to deer?" Ganz says. "That's going to influence how they interact with predators through a couple of mechanisms. One, how much are they limited by the available food? Two, how are they moving, relative to what food is available? We have all this GPS data from our collared deer and elk. We can look at it in areas where they have more exposure to predators. What sorts of habitats are they selecting?"

I ask her if the deer are more influenced by predators, or more by their habitat.

"When exposed to a lot of predators, they may be more selective about habitat structure and perceived predation risk," she tells me. "Whereas if predators were not a concern for that animal, you would expect them to choose their habitat based on food."

Her research may shed light on some of the thorniest wolf questions, all of which have real world and political implications. But today it's all about the tedious work of science—taking core samples from trees, determining how thick the underbrush is— which in turn tells us what kind of sight lines the predators and prey had during the deadly encounter. I help in whatever ways I can, and end up holding a plastic rod with bands of colored tape. Meanwhile, Ganz crouches near the site of the killing and tries to see what the deer saw, or didn't see, before her death. After taking core samples from the trees, Ganz determines that the stands are about twenty years old, more or less what she'd guessed.

While we're working, we hear the *wapp wapp* of a four-wheeler. A man drives up, dressed all in denim. With his silvery white hair and Fu Manchu mustache, he looks a little like a retired WWE wrestler who gave up the bright lights of Vegas for the quietude of rural Washington.

Ganz switches seamlessly from the role of scientist to the role of public relations specialist. She explains the project to the man, who points to a different thicket of trees and tells her that there is another deer carcass in there. He's friendly and curious.

But he's also checking up on us, making sure that we're not squatters or vandals. Although this isn't his land, it's obvious that he takes pride in keeping an eye on it. This reinforces my sense that, regardless of the results of Ganz's research, regardless of the science behind these deadly encounters between predators and prey, the people living on the ever-expanding edge of our communities have a stake in this land that's impossible to deny.

Chapter 7

PREDATORS
AND HUMANS

The crisp morning in early November started like many others, Craig Condron recalls. He was hunkered behind a scrum of logs and brush, rifle in hand, placed strategically along a well-worn game trail. With his head and shoulders barely visible above the blind, he'd already seen a doe and its mother wandering through the little clearing. Squirrels were now scampering through the woods. He'd been up for hours, an autumn ritual that he eagerly anticipates. A retired homebuilder, Condron has been hunting the same piece of land for fifty years.

It's dappled land, thickets of trees interspersed with pockets of more open brush. Logged many times, the trees are not large, and are still fighting to establish themselves. The longest clear shot Condron had on that morning was one hundred fifty yards. Now in full light, he was keeping an eye on the squirrels when suddenly he saw a flash of white out of the corner of his eye.

A deer, he thought at first. But as the name suggests, white-tailed deer have white tails, and the speck of white was moving toward him, rather than away. The speck exited the trees, resolving itself into the presence of a white wolf, followed in short order by five gray wolves. The pack picked their way out of the timber and onto the game trail running about fifty feet in front of Condron's hiding spot. They strolled by, mouths agape in that universal canine grin.

Condron had never seen a wolf, much less five of them, although for more than a decade he'd been seeing signs of their return—tracks in the snow and mud.

He'd also been seeing less tangible, harder to quantify signs, including some that were certainly not scientific. Condron had taught his son how to hunt on this land. Every fall, they'd bag a deer or elk, sometimes both. In fact, their success had been so assured that they took to passing up bucks, waiting for one that was "a little bigger," a common practice among hunters.

But starting about twenty years ago, and accelerating recently, it has become harder and harder for Condron and his son to fill a deer or elk tag. There aren't as many animals in the woods, he tells me, and those that remain are wary and hard to find.

Meanwhile, he's seeing more and more wolf and cougar tracks. Some days he sees only predator sign, without a huntable ungulate track in sight. Last year his son didn't get a deer after eight days of hard hunting. "It was sad," Condron says. Since the return of wolves, he's started carrying a .44 magnum (think *Dirty Harry*). "I'm always looking over my shoulder," he tells me. "Here I am, trying to enjoy the outdoors, and there is a little bit of fear factor."

So, when he saw the five wolves wandering out of the woods and into the clearing, he didn't feel any surge of fondness or awe.

The leader of the pack stopped. Sniffed. The wolf looked downhill toward Condron's hiding spot, even though he hadn't moved a muscle. He flicked his hunting rifle's safety off. The wolves got closer. Forty yards, thirty, twenty yards. *That's close enough,*

Condron remembers thinking. He stood up. Startled, the wolves scattered and then circled behind him, loping into the thick trees and bushes. Condron saw the white wolf nosing through the woods, so he stepped out of his hiding spot. This time the pack got the message. They disappeared into the forest, leaving Condron glancing nervously over his shoulder.

All of this occurred less than three hundred yards from the road—irrefutable proof to Condron, at least, that predators, and wolves in particular, are overrunning the woods. Unlike some folks, he doesn't hate wolves. He knows they're just doing their thing, providing for their families like humans do. The problem as he sees it is that they're just too wild for Washington State. "I guess they have a place in the world: the wilderness of Canada," he says. "I hate to say it, I don't know if they have a place in the modern populated world."

He doesn't want them to go extinct, but he also doesn't want them in his backyard, where he hunts and takes his grandkids for hikes. Although he doesn't use these words, it's clear from spending time with him that he feels as though he's on the front lines of the predator war—and, by extension, a larger culture war.

"I really enjoy nature," he says. "I can go up and hike in and sit on a rock and just take it all in." Just not wolves.

In the lower forty-eight, the farthest you can get from a road is about nineteen miles, and in most places it's much less than that. The United States is a nation of roads, and the Forest Service is our largest road manager. With about 380,000 miles of road under their purview, the Forest Service oversees roughly eight times more roadway than does the vaunted interstate highway system. The vast majority of these roads were built so logging companies could access remote and steep terrain.

The legacy of that government-subsidized industry is visible throughout the western United States, particularly when viewed from the air, when the checkerboard pattern of the land is revealed. This is a consequence of so-called "railroad grants," the government's practice of rewarding railroad companies with every other parcel of land along a rail corridor. The railroads, in turn, sold most of their holdings to timber companies.

Meanwhile, more than 80 percent of Americans live in a city or suburb, and that number is growing. The nation's century-long retreat from rural life has left formerly developed lands vacant, giving animals like wolves a place to call home. This shift is fundamentally changing how humans understand and interact with wildlife, which creates two problems—problems that are different sides of the same coin.

First, the concentration of people in urban settings creates "environmental amnesia," argues Aaron Wirsing, the UW professor. "You don't have any interest in protecting stuff that you're not even aware of," he says. In Washington, this means animals like wolves and bears get a lot of attention and money while even more endangered species are all but ignored. "Canada lynx, which are on the verge of being lost in Washington, are a nonissue," Wirsing points out.

This imbalance creates the second, thornier issue. For all the environmental amnesia, there are plenty of urbanites who are passionately committed to conservation—conservation of a few particularly charismatic species. "There is very little at stake for them," says Wirsing. But, on the flipside: "You have a diminishing rural population who are on the 'front lines' of predators. There is this massive policy skew."

But what does it really matter to me, a modern human living in a city far from chopped-up forestlands? Far from the lives and deaths of fawns, bears, wolves, and elk?

It matters. And just how much it matters has only become clearer over the past decade. Climate change and its increasingly evident impact is one thing. But a global pandemic, perhaps caused by the ongoing disruption of the earth's ecosystems, highlights the importance of understanding the tangled webs between predators, prey, and humans.

Consider a market, teeming with humans and animals: monkeys, pigs, bats, pangolins, beavers, rats, deer, and other animals for sale. Smashed into this small area of half a million square feet—slightly more than ten acres—one can find a kaleidoscope of species. These are animals that under other conditions would rarely, if ever, interact. But here they are being sold by butchers and grocers.

Or imagine an early October morning. The sun is rising, fog rolling off a distant ridge. A hunter has been up for hours in Wyoming's hill country and is now watching a monster bull elk saunter along the ridge one hundred fifty yards away. The hunter's heartbeat quickens. She raises her hunting rifle, the crosshairs hovering over the elk's front shoulder.

It's nearly silent, any sounds dampened by a dusting of snow. This is the largest elk the hunter has ever seen. The animal has

grown big and strong by nibbling the new growth that follows forest fires and logging operations. He's been mostly unbothered by
predators. In the cold, hard Wyoming winters, he eats feed put
out by wildlife managers. The hunter slows her breath and gently
squeezes the trigger.

On the surface, these situations have little in common. The first
is characteristic of the wet markets found in many parts of Asia.
These also happen to be the kinds of place where diseases can
jump the species barrier. Although the origin of COVID-19 is still
being investigated, one plausible scenario is that it originated in
bats, jumped to some as of yet unknown intermediate species, and
then mutated to infect humans.

Meanwhile, the superficially bucolic second situation actually
represents the front lines of a deadly neurological disease that has
slowly marched its way through elk and deer herds in the western
United States: chronic wasting disease, or CWD.

In the same way that these scenarios appear to be near oppo-
sites in tone and setting, the two diseases, CWD and COVID-19,
seem to share little in common—other than being poorly understood. One is a fast-acting virus that infects humans. The other,
a simmering neurological disease found in deer and elk. And yet
there are more similarities between them than one might imagine.

"Of course, COVID-19 and chronic wasting disease are very different diseases," says Margaret Wild. "But the ways by which the
diseases spread have some commonalities."

Wild would know. A professor at Washington State University, Wild has spent her career researching emerging infectious
diseases in wildlife. This has mostly meant studying CWD. (More

recently, she's been applying the knowledge and systems she learned while studying CWD to the poorly understood elk hoof disease that has increasingly impacted herds in Washington over the past two decades.)

CWD is a neurological disease that kills deer and elk. It belongs to a family of diseases known as transmissible spongiform enceph-alopathies. These brain disorders are caused by oddly shaped pro-teins called prions.

The effect is devastating. Infected animals stumble, drool, shed weight, and, among other things, lose all fear of humans. Then they die. For obvious reasons, CWD is often called the "zombie deer" disease.

Making matters worse, the infected animals shed prions, which can persist in the environment and remain infectious for at least two years, if not longer. In Colorado, where the disease was first documented in the 1960s, managers culled entire herds in hopes of stopping the disease. But when deer and elk returned to the landscape, lingering prions infected these healthy animals, and they, in turn, became carriers of the disease.

CWD has now been found in twenty-seven US states, three Canadian provinces, and the countries of Norway, South Korea, Finland, and Sweden. In 2021 it was detected for the first time in Idaho. The disease has yet to spread to Washington, although many believe it's simply a matter of time.

Thankfully, as far as anyone knows, CWD has never jumped to humans. This doesn't mean it can't happen; ongoing research into nonhuman primates, most notably in the case of eighteen unlucky macaques at the University of Calgary, indicates that CWD may infect monkeys. However, the results of this study have not yet been published, and other published research indicates that human prions are resistant to the abnormal folding observed in the disease. And yet, if it turns out that CWD can infect primates, it will be one step closer to humans.

Compounding this worry is the fact that another prion disease found in cattle did make the jump. Creutzfeldt-Jakob disease is the human variant of bovine spongiform encephalopathy, commonly known as "mad cow" disease. When confronted with such concerns, Margaret Wild is quick to emphasize that it's "not a given that [CWD] will make the leap."

But two years ago, the same could have been said of SARS-CoV-2.

The SARS-CoV-2 virus is in the coronavirus family of viruses. There are six other coronaviruses that are known to infect humans. Four of those are mild and cause roughly a third of all colds. The remaining two are rarer and more severe. They cause MERS (Middle East Respiratory Syndrome) and another strain of SARS (Severe Acute Respiratory Syndrome).

Meanwhile, there are at least five hundred identified coronaviruses found in Chinese bats.

Coronaviruses spread via respiratory droplets, not oddly formed proteins found in brain matter. And COVID-19, the disease caused by SARS-CoV-2, isn't nearly as deadly as most prion diseases. However, Wild points out, there *are* similarities in the ways in which these diseases spread, as well as the process by which they make the jump to humans—an event known as a spillover.

The coronavirus pandemic has thrust these concerns into the spotlight. Words and phrases once used only by epidemiologists, veterinarians, and public health officials—"flattening the curve," "social distancing," "surveillance"—are now common parts of the lexicon.

"Infectious disease is all around us," writes environmental journalist David Quammen in his book *Spillover*. "Infectious disease is a kind of natural mortar binding one creature to another, one species to another, within the elaborate biophysical edifices we call ecosystems."

Diseases, like animals, are trying to survive. In a sense, diseases are predators. But unlike large predators—wolves, for instance—they attack and eat their prey from the inside. Much as wolves do, diseases have their favored targets, species with which they co-evolved, honing their predatory skills over millions of years.

For most of life's history on Earth, evolutionary changes in diseases were slow. Incremental. But as landmasses shifted, or the occasional intrepid individual traveled a great distance, previously unacquainted species would meet and start sharing pathogens.

Diseases found new hosts. Hosts adapted.

Most efforts to jump species fail. Evolution is a brutal process, with many more dead ends than successes. But what it lacks for in finesse it makes up for in persistence. And humans have accelerated the rate at which things change, giving the evolutionary process more chances to get it right. As we've expanded our footprint—clearing forests, traveling the world, displacing animals—we've come into contact with long-sequestered pathogens.

"Lots of these disease problems are caused by stupid things that humans are doing to the planet," says Andrew Dobson, a professor at Princeton University. "It's really important for people to know that lots of the things we're doing to disturb natural habitats carry this hidden cost."

Dobson studies zoonotic diseases—that is, diseases infecting both animals and humans. Much of his work has focused on the Greater Yellowstone Ecosystem, a place where large wild animals are sharing terrain with nonnative domestic animals.

He argues that ecologists have traditionally ignored the role that pathogens play. And it's a big one: 90 percent of the Earth's biodiversity is parasitic, he points out. Most species have between ten and twenty parasitic species living inside them.

The U.S. Centers for Disease Control and Prevention estimate that 75 percent of new infectious diseases in humans originated in wildlife, and in the past fifty years, emerging zoonotic diseases have quadrupled. A group of more than one hundred conservation organizations emphasized these points to Congress in 2020. They urged lawmakers to include money in the stimulus package to address the underlying issues causing an increase in zoonotic diseases.

But Congress did not allocate funding, and as of 2021, President Biden had approved more gas drilling permits on public land than did President Trump, further weakening US regulations on the drilling, mining, and logging operations that are being carried out on public lands in the West. This is in spite of the fact that a 2015 study found that land use changes resulting from such projects have driven many of the zoonotic outbreaks of the last century.

Although it's unlikely that CWD will make the jump from elk and deer to cattle or humans, the more chances it's given to do so, the more likely it is to mutate and succeed. Imagine a gambler playing penny slots. If they play every day for a year, an unlikely event—hitting a jackpot—may happen.

Over the past one hundred years, elk and deer populations have grown in the lower forty-eight. It's been an incredible conservation success story. Meanwhile, more and more Americans have been building homes on land that had been agricultural. Attracted by the succulent yards and shrubs, elk, deer, and other animals are coming out of the hills looking for an easier life. With most of North America's native predators gone—or at least greatly reduced in number—the living has never been easier.

Wyoming even goes so far as to provide food for elk and deer in the winters in order to ensure better hunting opportunity in the fall (and to protect farmers' fields). Other states go to similar lengths, occasionally transporting them, whether for research or profit. All of this means that elk and deer are occupying a landscape that's mostly free of predators, living in herds managed by state agencies to maximize hunter success.

The bottom line is that these sorts of efforts have increased the possibility for a spillover to occur. CWD, after all, was first identified in captive mule deer herds.

"If CWD is a relatively rare disease, and people are infrequently exposed to it, then the likelihood of something bad happening is exceedingly rare," says Wild. "The more CWD we have in deer and elk, and the more people that are exposed to those animals, the more likely an exceedingly rare event could happen."

The strategies used to slow the spread of CWD are nearly the same as those deployed by public health officials hoping to flatten the COVID-19 curve.

When five cases of CWD were confirmed in Libby, Montana, wildlife officials in Idaho began undertaking year-round surveillance in the Panhandle. Idaho Fish and Game now collects samples from hunters and roadkill, and responds to reports of deer behaving abnormally. Their sampling model gives the agency a 95 percent chance of detecting CWD if 1 percent of the population is infected.

In 2018, the state updated its chronic wasting disease response plan, the third such plan Idaho Fish and Game has drafted. Additionally, last year the Idaho Fish and Game Commission banned the import of deer, elk, and moose carcasses or urine from areas with documented cases of CWD.

Washington banned deer farming in 1993, a decision that some believe has kept CWD out of the Evergreen State. State officials kicked off their own surveillance program in 2021. Melia DeVivo, a research scientist who focuses on ungulates for the Washington Department of Fish and Wildlife, has been working on the state's CWD response plan.

"Our risk is pretty high," DeVivo tells me. "It's hard to say what exactly that risk level is. But we are seeing a slow spread of the disease from state to state. We know we have animals moving naturally across borders. We also have risk posed whenever we move live animals."

"A deer in the wild is very rarely exposed to that prion," Margaret Wild adds. "But when we move an animal from across the country, alive or as a carcass, it's just like someone with COVID-19 flying into our city and infecting us locally."

Prevention is the key. Once CWD is established in an area, it's nearly impossible to eradicate. DeVivo knows of only one instance when wildlife managers were able to rid a landscape of CWD, and that was due to luck more than anything else. Instead, the best offense is an aggressive defense. This means removing attractants that cause deer or elk to congregate, whether these are baiting stations, salt licks, or feed lots, all of which "pose a huge risk of the transmission of the disease," DeVivo says.

Congregating animals pose an infectious disease risk, a fact that means Washington and Idaho may have a unique (albeit controversial) advantage in their battle against chronic wasting disease—a nearly full suite of native predators.

The two states are home to an impressive menagerie of wild animals: cougars, black bears, fishers, wolves, lynx, and even the occasional grizzly. And when it comes to CWD, these predators could be the entire region's saving grace.

"Predators do a very good job of removing sick animals from populations," Andrew Dobson says. "And that reduces the rate of transmission."

DeVivo's research indicates that predators, particularly wolves and cougars, can naturally constrain the spread of wildlife diseases. "Cougars actually seem to selectively prey on CWD-positive deer," she says.

But wolves might actually be the best bet because of how they hunt. They chase their prey, testing herds of deer and elk, hoping to isolate the weakest and slowest, which would likely mean CWD-infected animals. DeVivo is quick to point out that the CWD research isn't conclusive, and as of now it remains a theory.

"We don't have a good example at this point of what the disease will do in a population where you do have a full suite of predators," she says. "I'm not going to say that it's going to be our saving grace for this disease. But I am going to tell you that it will be very interesting if CWD does show up in northeast Washington."

Back in Wyoming, the hunter hits the bull elk. A clean shot. She spends the next twelve hours field dressing the animal and hiking it back to the car. It's only five miles, but she's hauling hundreds of pounds of meat, after all.

It's dark when the job is finished. The hunter knows that Washington prohibits the transportation of the heads of elk and deer into the state from CWD-positive places like Wyoming (unless all the soft tissue has been removed), but it's late, the hunter is tired, and this animal is as healthy-looking as they come.

She drives through the night, the elk's head and antlers wrapped up and tucked away.

Nothing happens.

The elk didn't have CWD. Or maybe it did, but for whatever reason the prions didn't linger. Washington stays CWD-free and the disease doesn't jump species.

This is the likely outcome.

Now, scale it up. Imagine thousands of hunters transporting thousands of carcasses. Each interaction representing a chance, however small, of something catastrophic happening.

There is no simple answer for how to deal with this scenario. No single solution.

"Really, health is about the health of all species," says Wild, the WSU professor. "By promoting the health of all species, and the environment, that's how we can keep ourselves the most healthy."

Chapter 8

WHAT OF
THE WOLF?

A flash of shaggy black fur. The hindquarters of a canine melt into the forest. We've just rounded the corner of a remote Forest Service road in north Idaho. I slam on the brakes and turn off the car, my climbing partner and I staring greedily into the underbrush. What was that?

I walk into the woods, trying to move slowly, but in my excitement I can't keep from moving fast. About ten feet off the road, right before the ground gives way to a steep hill, I stop and scan the land ahead.

And there it is. A wolf. A wild wolf, about one hundred yards downhill from me, looking back up the hill. We make eye contact. I study her black coat. Her pointed snout. She's smaller than a husky and lean. I've decided the wolf is female.

We stare at each other for what feels like a minute or more—likely less—and then she turns and trots off, leaving me at a loss for words. But the eye contact has offered me a sense of a profound otherness in this creature that's living by her own code, moving through the world with her own agenda.

There are plenty of wild animals you could say this about. So what is it that feels so different about getting stared at by a wolf? There's something familial going on here, a whiff of our species' shared history.

Until this moment, I've been baffled by the passions that wolves incite. I do not understand the utter hatred professed by some for these creatures. Nor can I relate to the passionate love, verging on obsession, exhibited by others. How can you feel so strongly, either way, about a creature you'll be lucky to see once in your life?

But standing here in the woods of north Idaho, staring after her, I feel something of these passions. I realize that I'm shaking. My heart is racing.

Just minutes ago I was driving through this world, ignorant of the lives and dramas unfolding just off the road. Now, everything around me is laden with hidden meanings. My senses are heightened. The woods have become vivid.

Over the following days, as I stew in the aftereffects of this encounter, I come to the following conclusion, a conclusion that is hardly mine to claim: The wolf is a reminder. A reminder that humans are not the only creatures with desires and wills.

This can be a disturbing realization. It challenges our dominion of nature. After an earthquake, our confidence in the immutability of rock and stone and earth and water is shaken. We walk gingerly, at least for a while, upon the crust of this planet, aware suddenly of our own mortality. A wolf's gaze can have a similarly seismic impact. A wolf's gaze, impassive and independent, challenges our desires. It disabuses us of the notion that we bipedal apes are the sole inheritors of this planet.

This, I believe, is why wolves incite such strong emotions. At the end of the day, it's not really about wolves. It's about humans, and our own unique blend of fears, desires, and hopes.

Consider the rancher. A hardworking person, undertaking an ancient and biblically ordained profession: raising livestock. For the small- to medium-sized operator, this is a calling as much as a business. A way of life.

It has fallen out of favor.

The big feedlots have depressed the price of beef, transforming the industry model into something more akin to a factory than to a farm. Other countries, with more relaxed environmental regulations and more exploitative labor laws, have taken business. At the same time, the culture has changed. A country once steeped in the mythos of the cowboy is reckoning with the violent and racist underpinnings of this idolatry, and acknowledging our brutal history of stolen lands and genocide.

Added to this is the growing interest, especially among urbanites, of restoring ecosystems and returning predators to the landscape. Where does this leave the rancher? Their parents and grandparents spent lifetimes building their legacies. Now it is threatened. A rancher staring into the eyes of a wolf would probably feel all of these threats coalescing into this one very specific encounter, with this one very specific creature.

And now consider the environmentalist. Statistically speaking they live in a city, or a suburb. They have little daily interaction with the natural world and no reliance upon it. On the weekends the environmentalist escapes the city and heads to the mountains and feels relaxed and whole.

In their eyes, nature represents something pure. A state of existence removed from the stresses of human life, particularly urban human life. It's only natural, then, that environmentalists ascribe values that they wish to see in themselves onto the natural world: harmony and connection, devotion and balance. The "natural order of things" becomes a stand-in for their own desire for community and meaning. And what represents this better than the fiercely social wolf?

But what of the wolf? What does she feel, think, and need? What are her hidden desires?

Scientists, biologists, ecologists, and wildlife managers have typically frowned upon this line of inquiry. Crediting human emotions and feeling and thoughts to nonhuman animals is not good science, they say. Anthropomorphism is a nasty insult leveled against those who dare see animals as individuals, not populations.

The irony, at least in the case of the wolf, is that this approach— of thinking about wolves as populations, rather than individuals—has caused nothing but grief for all involved.

When I connect with him, Francisco J. Santiago-Ávila is animated and thoughtful. A PhD student at the Carnivore Coexistence Lab at the University of Wisconsin-Madison, he is also the proponent of some controversial views, primarily this one: Why not consider the individual animal when making wildlife policy, instead of leaning on abstracted and lifeless terms such as "population?"

"What you're seeing is that a lot of agencies, once the numbers look good, they don't care," he tells me. "There are no questions about wolf society. There are no questions about wolf stability. It might be the case that you can just keep killing individuals sustainably, but you're destroying a whole society."

This may sound a little *woo woo* when compared to the hard-boiled facts and figures approach of science. But Santiago-Ávila pushes back against that way of thinking.

"There aren't any questions being asked about what wolves are experiencing," he says. "I'm being taken to task for it. Professionally, as a conservationist, it's still very hard to have conversations about more than just numbers."

Taking such a view is seen as being nonscientific, even biased. More than a few of his papers have been rejected by prestigious journals. Fortunately for Santiago-Ávila, he's backed up by Adrian Treves, his PhD advisor. And emerging research has highlighted the need to examine both the macro and micro levels when making policy decisions.

Most recently, the two researchers examined the impact that liberalized periods of state-sanctioned wolf killings ends up having on poaching. It's an interesting question, one that challenges a core rationale for killing wolves and other predators, namely that doing so increases social tolerance. This line of reasoning makes some intuitive sense: If wolves attack cattle, or other livestock, and the state kills some wolves, then everyone will be happy. Tolerance for the animals increases.

However, Treves and Santiago-Ávila's research indicates that this is not what's happening. Instead, in a broad review of wolf mortality data from between 1979 and 2012, the duo found that the number of wolves that disappeared had increased during periods of liberalized state killing. The duo hypothesized that this increase in missing wolves—categorized as Lost to follow-up, or

LTF—is due to so-called "cryptic poaching," an oxymoronic term worth considering.

Coined by Swedish researchers Olof Liberg and Guillaume Chapron in 2011, cryptic poaching is defined as all the deaths that can't be directly attributed to poaching, but are in all likelihood the result of poaching.

By its very nature, poaching is a clandestine activity, one that is not reported or tabulated (unless the poacher is either caught in the act or caught with evidence of the act). This means that a large percentage of wolf deaths are simply mysteries, LTFs. One Scandinavian study estimated that half of all wolf deaths were due to poaching and two-thirds of those deaths were deemed cryptic.

Most agencies, and many researchers, have never really looked closely at this category. Instead they've discarded "the ones that were lost," in Santiago-Ávila's words. When faced with LTFs, researchers have typically extrapolated from the known causes of mortality, including confirmed poaching, applying this to the entire category.

To some degree this does make sense. Attempting to infer about what might have happened to an animal without evidence of what actually happened can be a risky game for researchers, particularly when it's connected to a topic as contentious as wolf recovery. But the lack of examination of the LTF category itself leaves a gaping hole.

The Swedish team's 2011 paper, provocatively titled "Shoot, shovel and shut up: cryptic poaching slows restoration of a large carnivore in Europe," tackled the problem in a different way. Liberg, Chapron, and the other authors didn't shy away from addressing the problem of inference early in the study. "Here we have demonstrated a high incidence of poaching in a threatened wolf population," they wrote, "but because a major part of this poaching was unobserved (cryptic) and inferred from indirect data, its estimate is open to criticism."

They determined that more than half of all Swedish wolf deaths were due to cryptic poaching, and without this additional source of death, the Swedish wolf population would be four times as abundant.

Treves and Santiago-Ávila picked up this line of inquiry. In their review of the available mortality data in Wisconsin, they found an increase of Lost to follow-ups during times of liberalized legal wolf killing. Santiago-Ávila argues that this increase in LTFs correlates with social science research they've conducted, in which they've found that certain types of people, hunters primarily, feel more inclined to poach wolves during liberalized periods. If their research is correct, wolves are experiencing much higher levels of cryptic poaching than previously imagined, anywhere between 11 and 34 percent.

"To come to a better conclusion of what's going on here," says Santiago-Ávila, "we would like better data."

If successful, his efforts to center research around individual wolves may have a say in what happens in Washington's ever-simmering Wolf War. His work also points toward a larger, more radical and altogether more difficult endeavor. That is, changing the very way we see and interact with the nonhuman world.

Human-wildlife coexistence is a hot topic now in ecological circles. How can we, this globetrotting, landscape-changing species, learn to live alongside wild animals? There are more people on the earth than ever before. And simultaneously there's less space.

For decades, if not centuries, coexistence in North America has meant relegating wild animals to places that are far, far away from where people live. This was the impetus behind the creation of the National Parks system. Protecting the "last best place." Sure, we humans took the valleys. We took the fertile plains and gentle forests, the meandering rivers and pristine bays alongside the oceans. Wildlife? They can have the other spaces—the mountaintops and the wild interiors.

But then we slowly filled in the wild interiors. The mountaintops, while severely beautiful, do not make the best habitat, as any benighted mountaineer well knows. And with the reality of climate change becoming clearer every year, even these remaining pockets of wildernesses are impacted by humans as surely as is a road-stuffed acre of Forest Service land.

Environmental exploitation seems to follow a sadly predictable track. A new, seemingly endless resource is discovered or accessed (North America's forests, for example). That resource is then logged, mined, or excavated nearly to death—in return, supporting a huge growth in human comfort and material safety. And then, moments before it's too late, we start worrying about losing it.

At the time of this writing, Canada continues to log old-growth forests and mine distant mountaintops, using the resulting economic prosperity to support a bevy of humanistic programs in healthcare, among other fields. This pattern has repeated throughout human history. Europe deforested, the Middle East deforested, rainforests logged, lithium mined.

Are we doomed to repeat this cycle until there is no more repeating? Until we are no longer able to push the concerns of the natural world over one more hill, out of sight for one more decade? Perhaps. But there is an undercurrent of radical change that may offer hope.

The notion that humans have inherent dignity as individuals is baked into much of our culture. Although there are certainly governments and cultures that reject this, it's more the norm than not. But this is a new development and hardly a preordained one.

In the same way, if we are to learn to live with and alongside the natural world, and if we are to preserve what is left, there must be a similar shift in view. A similar revolution. These ideas may sound kooky when plugged into the dry and rational world of science and bureaucracy. The dignity and rights of animals? The experience of the individual wolf? Perhaps these sorts of concerns sound more like personal choices or philosophies than they do science or management techniques. Increasingly, however, experts are making the argument that this is not the case. That this shift in focus will be key to the ongoing survival of both humans and our wild brethren.

Stories of wolf attacks, of wolf killings, come up again and again over the course of recorded history. If we are considering the wolf as an animal with inherent dignity and rights, then we must consider this side of her story, too.

In October 2018, a twenty-nine-year-old man from New Mexico named Jordan Grider headed into Minnesota's Boundary Waters

Canoe Area Wilderness. Grider's self-proclaimed mission, which he shared with his mother before he left, was to spend the winter living "off the grid" in the nearly 1.1-million-acre wilderness, where he planned on being entirely self-sufficient.

"He had kind of given me the directions," his mother Rebecca Grider tells me. "But I didn't think to pay much attention to it."

By Christmas, when her son hadn't yet checked in with her, she began to wish she'd paid better attention to his plans, although it was not unusual for her son, who had a penchant for adventure, to disappear into the woods for long stretches at a time. Seven months later, as the deep freeze of the Minnesotan winter loosened its grip, game wardens and police trekked into his camp spot. There they found a macabre mystery.

The camp was still up, with Grider's hammock slung under a green tarp—everything still hanging off a guy wire. They also found a 9-millimeter Beretta along with a few days' worth of beans, rice, and wheat. Inside the tarp they found a blood-soaked sleeping bag and wolf scat and tracks. But no Jordan Grider.

He was always different, his mother tells me. Born and raised in the small agricultural town of Moriarty, New Mexico, just east of Albuquerque, he was one of six boys. From an early age it was clear that he was dyslexic, his mother says. When he was five years old, specialists showed him a series of photos, first a dirty dog, then a dog in a bathtub, then another photo of a dirty dog. Jordan was supposed to line these photos up in a sequence that showed that "the dog was dirty, got clean, and then got dirty again," says Rebecca Grider. "He couldn't handle that. He was five."

Rebecca Grider is a religious woman who homeschooled her children before going back to school and becoming a medical massage therapist. She and her husband run a telecommunications business called Western Tel-Com, Inc, or WTCI. The W of that acronym represents their mission to "Witness and demonstrate Christ's love for people in everything we do and say, and to uphold Godly values."

Two years after her son's death, she speaks about him easily and clearly with some hard-earned practice. Talking about him is obviously therapeutic for her. Plus, she's writing a book about his life, an ode to his adventurous spirit as well as an examination of his mysterious death. Sort of like *Into the Wild*, although she's quick to emphasize that Jordan Grider was a more experienced outdoorsman than Christopher McCandless.

Despite his dyslexia, or perhaps because of it, he was a creative and inventive boy who grew into a curious, adventurous man. His mother remembers one of his teachers posing young Grider a question: "If you call someone who teaches a teacher, what do you call someone who sings?"

"A fat lady," he deadpanned.

He was homeschooled for most of his education, only attending the local high school for his junior and senior years. "He was gifted and yet challenged," his mother recalls. "He got in a lot of trouble in high school."

When he was eighteen, he started working at a nearby truck stop and started hanging with a rougher crowd, doing drugs and drinking. That led him to New York, near Rochester. As his mother tells it, one night, high out of his mind, he saw the angel of death coming for him. In short order he got clean, found Jesus, and started spending more and more time outside.

Over the next decade he bounced between New York, New Mexico, and Kentucky. He built furniture for Mennonites, worked at Walmart, and built barrels for Kentucky bourbon. He spent much of that time living outside.

Then, in 2018, after splitting with his girlfriend, he came back home and lived with his parents for six months before departing for Minnesota in October. His parents tried to convince him to stay closer to home. Why not the Rocky Mountains in New Mexico? they asked. But Grider's mind was made up.

The last text Rebecca Grider got from her son was on October tenth, 2018. It was a photo of a lake, and some bags of beans and rice. "I found my new home," he wrote.

Several days later, a conservation officer from the Minnesota Department of Natural Resources named Sean Williams received a complaint that a pickup was blocking a private gate. Williams and a US Border Patrol agent responded, wondering if perhaps someone was sneaking across the border. They found Grider's truck but no sign of him. So they called his mother. Eventually she paid for the truck to be towed out from in front of the private gate and down the road.

She was worried, but not panicked. By Christmas, with the snow piling up in Minnesota, Rebecca Grider knew something bad had happened, although she still held out hope that her son would be rescued.

In April, as the weather improved, Sean Williams headed back out to look for Grider. That's when he came across the bloody campsite.

On a subsequent trip, after more snow melted, Williams found what remained of Jordan Grider: just twelve bones, including a vertebra, femur, and forearm. Subsequent DNA analysis confirmed that these were Grider's bones.

The lanky redhead with buzzed hair was dead at twenty-nine. But how?

Investigators have been quick to emphasize that they don't have any definitive answers to this question. "Honestly, I'm sure we'll never know exactly what happened," Williams said in one news story about the death. "I sort of lean toward he had some sort of accident and cut himself or stabbed himself, something like that."

Rebecca Grider is pretty confident she knows what happened to her son.

"He probably got attacked by the wolves that night," she says. She thinks that after he went to sleep, the wolves ambushed him. This, she believes, is why there was no struggle.

In 2019, a wolf attacked a family camping in Canada's Banff National Park. The animal shredded the tent and was in the process of dragging a full-grown man out of the tent when a nearby camper chased the animal away.

Afterward, in news story after news story, experts cautioned that this was a "very rare incident." The standard line is that wolves rarely attack humans. Looking at the recent history of the West, this seems to be true. According to a 2002 report investigating eighty wolf-human encounters in which wolves showed little fear of people in Alaska and Canada over the course of six decades, healthy wolves bit people or their clothes sixteen times. None of these attacks were life-threatening.

A review of global wolf attacks, also published in 2002, found that "there are no documented cases of people being killed in predatory attacks by wolves in North America during the twentieth century."

And yet there are stories of wolves attacking humans throughout the history of the American West.

Another study examining the commonalities between wolf attacks worldwide concluded that there are four primary conditions that lead to more frequent attacks: rabies; habituation to human presence; provocation (such as stumbling upon a wolf denning site); and highly modified environments.

The authors of the review also found that the majority of predatory attacks in pre-twentieth century Europe and present-day India have occurred in "very artificial environments." In those environments, five commonalities emerged:

- little or no natural prey
- heavy use of garbage and livestock as food by wolves
- children unattended or used as shepherds
- poverty among humans
- limited availability of weapons

"The results of this report may seem surprising to many, given the modern positive image of the wolf as an almost harmless carnivore," the study concluded. "The main symbolic conclusion that comes from this study is that it is time to stop viewing the wolf as a devil or a god. A wolf is a wolf. As a species we cannot expect them to not eat humans (an easy and abundant prey) on principle. We should just be glad that they avoid us as much as they do and manage them to keep it that way."

The view that wolves essentially pose no danger to humans has, at times, verged upon dogma. Following the Canadian wolf attack, the International Wolf Center released the following statement: "A person in wolf country has a greater chance of being killed by a dog, lightning, a bee sting or a car collision with a deer than being injured by a wolf."

On the flipside, stories like Jordan Grider's and that of the family in Canada are championed by anti-wolf advocates as proof that wolves are a menace to people, and that the animals should be aggressively managed. Following the attack in Canada, a Washington cattleman shared a story about it to his Facebook page. "Coming to a Stevens County campground near you," he wrote.

"Those with confidence in scientific knowledge are likely to be more positive towards wolves," the 2002 review of wolf attacks concluded. "Since scientific knowledge holds a hegemonic position to lay knowledge, the contestation of claims that wolves are harmless may be an element in a struggle against the dominance of this form of knowledge."

Valerius Geist is an unlikely champion of the anti-wolf movement. Now in his eighties, he is a pioneer in the study of mountain sheep, deer, moose, and other North American ungulates. An influential researcher whose methodology was decidedly old school (he officially retired in 1994), he's spent months in the Canadian wilderness, has written or edited more than fifteen books, loves hunting and fishing, and has regularly served as an expert witness in wildlife-related cases and legislative deliberations. He's also an erudite and effortlessly charming conversationalist with a penchant for moving nouns to the front ("God thanks") and rolling his R's—a legacy of his mother tongue, German. A lifelong hunter and angler, his voice cracks when describing the beauty and intelligence of a moose. He cries on a podcast when speaking of his wife, Renate, a talented biologist herself, who died in 2014.

In the 1990s, Geist fought against game farming despite having previously supported the practice. His earlier support had been a misguided attempt at figuring out a way to make wildlife pay for itself; he'd quickly realized that game ranching would spread disease, which would ultimately lead to the privatization of wildlife. This change of opinion was met with disdain and vitriol from the nascent elk ranching industry. Geist received death threats, as did his wife. After presenting "one of the most innocent papers" at a conference in Texas, one of his Texan colleagues pulled him aside and said, "We're all with you, but we can't be seen with you." Geist's paper had been about why Irish elk have such big racks.

Audubon magazine dubbed him "the man elk ranchers love to hate." One rancher called Geist an "alien from another planet." "The man is crazy," another rancher was quoted as saying. "The man is a radical." He was the topic of at least one panel discussion at the North American Elk Breeder's Association in 1992.

His crime? Arguing that game farming would lead to the outbreak of wildlife diseases. The thing is, he was right. The spread of chronic wasting disease throughout the United States has been fueled, in part, by game farms and wild ungulate transport.

"It's not a joy to be right," Geist said in a 2004 interview. "I only asked people to do their homework, which the government didn't do. Unfortunately, this is becoming a feature of North America—how much knowledge there is and how little governments pay attention to it."

Now a widower approaching the end of his life, Geist finds himself delivering yet another alarming message, except this time he's not being attacked by ranchers and hunters for his conclusions. Instead, he's allied himself with them in warning against the excesses and dangers of wolves.

"I'm most embarrassed to say that I'm a professor emeritus of environmental science," he tells me by phone. "Environmental scientists have lost their passion for knowledge and truth. They have sort of a conviction that human beings are an evil source that controls the world, and the best thing we can do is let nature take over."

Geist was born in 1938 in the breezy port city of Mykolaiv, then part of the Soviet Union, now Ukraine. A hub for shipbuilding since its founding in 1789, the city is located forty kilometers south of the Black Sea, on the Southern Bug River. Geist's mother and father were engineers, of submarines and battleships respectively. They left the Soviet Union when Geist was four years old, moving first to Austria, and then to Germany, where Geist became, in his words, "culturally a German."

In 1953, when he was fifteen years old, the family moved to Canada. Straightaway he joined the Regina Rifle Regiment (now the Royal Regina Rifles) for a simple reason. "I had enough experience on the receiving side in World War II," he says.

Geist has continued working well past his formal retirement in the 1990s, and has long been a champion for public land and public wildlife. In 2019, he helped edit *The North American Model of Wildlife Conservation*.

His lightbulb moment, at least when it came to wolves, happened just before the turn of the century, several years after he'd retired and retreated to an agricultural area on Vancouver Island. Before settling on the island, he'd primarily studied mountain sheep and other large-bodied North American mammals, and was one of the founding architects of the North American Model of Conservation.

This model—which wasn't named until 2001, but was developed over the course of approximately one hundred fifty years—has been credited with preserving, and, in some cases, saving much of North America's native wildlife. A key tenet of the model is to pursue the active engagement of hunters and anglers, who, through licensing fees, advocacy, and taxes, fund conservation.

The North American Model of Conservation is considered by many to be the premier conservation model on the planet, although it's not without its detractors. The critics claim that, due to the funding structure, hunters and anglers end up having an undue influence on how and what species are valued. This dynamic largely explains why fish and game agencies have been just that: fish and game-focused.

Before moving to Vancouver Island, Valerius Geist had tended to believe what others told him about wolves: namely that they weren't dangerous to humans, and that they could and would strike a balance once reintroduced.

He'd also encountered wolves while doing fieldwork. In a blog, he describes "exceptional opportunities to observe wolves in pristine wilderness" while studying sheep in northern British Columbia. "My closest neighbors, a trapper family, lived some forty miles to the west, and the closest settlement of Telegraph Creek was about eighty miles to the north," he writes. "Timberlines were low, and the wolves spent much time in the open, plainly visible. I watched them for hours on end."

The wolves Geist saw were large because of the abundance of game. They were "painfully shy" because of the abundance of guns, and because they had sustained a century of human assault. "They even panicked over my scent," he writes. "Though they killed a few sheep, their hunts were largely unsuccessful. However, I began to appreciate their strategies and tenacity as hunters. In traversing the valley, I crossed a wolf track about every fifty paces. They were that thorough in scouring the valley for moose."

His opinions about wolves began to change once he moved to Vancouver Island.

"I believed in those days that what my colleagues were doing was probably right," he tells me. "But what I realized later on was they had judiciously avoided reading history. They were very proud men."

Vancouver Island is so large that, if one were to zoom out far enough on Google Earth, it would look contiguous with the West Coast of Canada. However, the island is actually separated from the mainland by the sixty-eight-mile-long Johnstone Strait, a saltwater channel that's approximately two to three miles wide.

As is the case with most islands, the native wildlife on Vancouver Island is particularly susceptible to extinction, isolated as these animals are from larger source populations. The island once had a genetically distinct wolf population, but government-sponsored trappers and hunters killed off those wolves in the early twentieth century.

However, wolves from the British Columbia mainland continued to swim across the Johnstone Strait, attempting to recolonize the island, and they continued to be killed until laws and mores changed. Showing off their species' amazing adaptability, these island wolves diversified their diet, chomping down waterfowl, otters, salmon, seals, and even occasionally sea lions. Finally, in the 1970s, they reestablished themselves on the island.

At the same time that wolves were reclaiming their territory on the island, so were humans. In the 1970s, nearly half a million people lived there. Now, more than eight hundred thousand do. Although that kind of population growth isn't unique, the pressures of an expanding human population are starkly felt in the nearly closed ecosystem that is an island.

Still, when Geist moved there in the 1990s, he landed in a rural area, replete with wildlife. On his long walks he became accustomed to spotting over one hundred black-tailed deer and six large male black bears. As many as eighty trumpeter swans would overwinter near his home, along with large flocks of Canadian geese.

This pattern continued, more or less uninterrupted, until January 1999. That's when Geist and his eldest son cut the tracks of a pair of wolves. By summer, a pack had established itself near his retirement getaway. "Within three months not a deer was to be seen, or tracked, in these meadows—even during the rut," Geist writes. "Using powerful lights, we saw deer at night huddling against barns and houses where deer had not been seen previously. For the first time deer moved into our garden and around our house, and the damage to our fruit trees and roses skyrocketed."

Geist observed behavioral changes among the other species he had grown accustomed to seeing around his new home. "The trumpeter swans left, not to return for four years, until the last of the pack was killed," he writes. "The geese and ducks avoided the outer meadows and lived only close to the barns. Pheasants and ruffed grouse vanished. The landscape looked empty, as if vacuumed of wildlife."

Over the first twenty-five years after they returned to Vancouver Island, there were less than a dozen recorded wolf sightings in Pacific Rim National Park, the rugged preserve located on the island's west coast. During the summer of 1997, two wolves followed a woman for half an hour during her evening stroll on the beach. This was such an anomaly that the event garnered plenty of local media attention.

But this was only the beginning. By 2003, there had been fifty-one recorded encounters between wolves and humans. To make matters worse, wolves killed seven dogs and injured one person during those six years.

"We are in a new wolf era," announced Bob Hansen, Pacific Rim's human-wildlife conflict specialist at the time. Hansen's warning turned out to be prophetic.

Since then, wolf attacks or encounters of one kind or another happen every year. Unleashed dogs are a favorite target, and in 2017 the park recorded its first leashed dog attack. Humans have also been injured. It's becoming more and more apparent that these island wolves are no longer behaving like the rest of their North American relatives.

Geist worries that what's happened on Vancouver Island will happen in Europe and North America. He's become something of a pied piper for the anti-wolf lobby. He testifies in court cases and speaks to hunting groups. In an email to me, he calls the wolf reintroductions "unconscionable, and thoroughly unethical to boot."

His theory as to why this all happened is simple and intuitive in its own way.

"Wolves were the lucky, lucky survivors of the Pleistocene extinctions," he says in one video. With most other large predators dead, Geist believes that wolves were left with no natural predators aside from humans. "In other words," he continues, "what we're looking at today is wolves out of control. Out of natural control. Completely out of control because their natural enemies have died out."

In a federal court case filed in eastern Washington, Geist testified that "large-bodied wildlife was very scarce not only in the South but also in the pre-Columbian West." Wolves (and presumably other large wildlife) only "became common when the buffalo population exploded after the massive die off of Native people. . . . Estimates vary, but some 56 to 120 million people died, upon which wildlife proliferated, generating conditions of super abundance, which environmentalists falsely believe were pre-contact conditions."

Geist believes that the modern environmental movement has built its core precepts around a hollow façade, a verdant paradise full of wild game, mostly unmolested by humans. He's not alone in arguing that our presuppositions about the natural world have been based on false information.

It's been shown again and again that the North American landscape that European colonizers set their eyes on—the landscape immortalized in paintings, songs, and writings—was the result of a viral holocaust that afflicted the Indigenous population. "As soon as the heavy hand of human beings came off the planet, wildlife exploded," Geist tells me.

And Geist is right, even if some of his conclusions about what this could signify for wolf management may be shaky.

In the 1960s, the tenor and direction of National Park management was greatly influenced by the so-called Leopold Report. The nineteen-page report argued that a primary goal of the National Park System should be to preserve the parks "in the condition that prevailed when the area was first visited by the white man. A National Park should represent a vignette of primitive America."

Ultimately, this philosophy still girds the environmental movement. In 1999, professor and historian Dan Flores published an essay examining the belief that "'the model natural ecosystem' is the 'pre-settlement' one."

Flores kicked off his 1999 essay with a quote from Reed Noss, a prominent ecologist who helped develop the field of conservation biology. "For any landscape," Noss wrote in 1983, "the model natural ecosystem complex is the pre-settlement vegetation and associated biotic and abiotic elements." Flores picks apart this view as a relic of an old ideology, one that needs to change. The time had come, he wrote, for an "intellectual paradigm shift for the descendants of colonizing Europeans across the world."

Think about the enormous herds of bison that white settlers encountered as they pushed west. These herds were most likely the combination of two things: first, the massive die-off of large-bodied carnivores during the Pleistocene extinction, and second, the intentional management that the Native tribes of the Great Plains undertook in their efforts to preserve an important food source.

Accounts from the Lewis and Clark expedition bolster this argument. In their journals the explorers puzzled over the huge difference between the prevalence of wildlife they observed living along the Missouri River, as compared to the Columbia River. Later research indicated that the ubiquity of game along the Missouri was due to the existence of a large "no man's land" between warring Native tribes. On the other hand, the relative dearth of wildlife they saw along the Columbia was likely due to the relatively peaceful, prosperous, and numerous tribes residing in the area.

"I have observed that in the country between the nations which are at war with each other the greatest numbers of wild animals are to be found," William Clark concluded in his journal. The indication, it would seem, is that peaceful, prosperous, and happy humans are bad news for wildlife.

When it comes to wolf research, Geist believes that his former colleagues have discounted history and unfairly cherry-picked the data with which they do their work. His skepticism didn't develop overnight.

Geist's central piece of evidence for this theory is the narrative of Álvar Núñez Cabeza de Vaca, a Spaniard who traveled from modern-day Florida to Mexico between 1527 and 1536. During those years, in which he alternated between being a guest, captive, slave, trader, surgeon, and eventually a prophet to the Native tribes he encountered, he documented a region thick with people and conspicuously empty of wildlife, all before European-introduced diseases devastated the Native populations.

Cabeza de Vaca has been praised by historians for his sober, methodical notations. As a nobleman who loved to hunt, he maintained a keen awareness of wildlife, and, as Geist points out, he was writing for an aristocratic readership that loved nothing more than a good chase.

"Thus Cabeza de Vaca saw the distinction between rabbits and hares, as he would have customarily done in Spain," Geist writes, "and he lists species of hawks (marsh hawk, sparrow hawk, goshawk) and falcons, as well as a favorite quarry of European nobility in falconry, namely the herons. Consequently, his account of herons, egrets and ibis. He mentions no eagles, which is significant, as that's a bird beloved by European royalty."

When he does describe novel fauna, Cabeza de Vaca is remarkably accurate.

"He reports three kind of deer in Florida," Geist continues, "and there are indeed three subspecies of white-tail deer in that state, but they differ little except in size. This suggests keen power of observation as well as substantial hunting experience. He makes a point of describing to the emperor wildlife not seen in Spain, i.e. the opossum."

Viewed through this lens, it's notable what Cabeza de Vaca did not report. He mentioned no elk, despite the likelihood that, as Geist suggests, "he would have commented on the great similarity to the red deer of Europe, as did later travelers that saw the elk." In all his years of traveling through modern-day Florida and Louisiana, Cabeza de Vaca also did not see a single alligator.

Instead, he reported widespread hunger (and nearly starved to death himself). Geist pays particular attention to the scarcity of food:

> Note the observations that Native people did not eat every day, that food was often scarce and hunger common, that there were rituals to make hunger bearable, that Álvar himself lived that very hunger for years as a captive, slave, or guest of Natives, and that children were being nursed into their teenage years, which the natives themselves ascribed to sporadic food availability.

Geist is quick to say that he's not anti-wolf. Instead, he believes that wolves need to be strictly managed—in other words, killed—to keep them from decimating game populations and hurting people. "Wolves mustn't be coddled if we hope to balance them with modern ecosystems—and to avoid becoming prey," he says. He's become something of a hero for ranchers and hunters who see in his conclusions a kind of validation for their experiences with wolves.

Geist's conclusions are very nearly antithetical to those of a researcher like Francisco Santiago-Ávila.

But the two camps do share some similarities. Geist and Santiago-Ávila both agree that most ecologists and biologists are divorced from the reality of nature. "There are people who are out in the field, and they are being ignored," Geist tells me. "We have the false belief that just because you're a computer whiz you have the truth by the tail."

One winter day, while Daniel Curry and I are trekking through the snow looking for wolf sign, he begins to explain to me how best to inhabit the mentality of a wolf. Think about the landscape, he says. Where is it easiest to walk, where is it not? Then think about where the prey is or isn't.

Curry's view of the world, and wolf management, centers the wolf and its experience, while also setting firm and clear boundaries. That's why, whenever he encounters a wolf, he does everything he can to scare the heck out of it. To reinforce that humans, and our concerns, are not good to be around if you're a wolf. Of course, Curry is pro-wolf, and scoffs at the idea that wolves are dangerous to humans, or that they will somehow overwhelm prey populations. But he's also a realist about what wolves are and what they are not.

"If I went outside, and there was a wolf chewing on [one of my horses], I would shoot that wolf," he tells me. "I wouldn't even think twice. I'd go grab my rifle and I'd go out there and shoot it, with the caveat of having done everything prior to that to prevent that meeting from ever happening."

In contrast, the reductive, dichotomized way of thinking that allows most of us to disassociate from the natural world is the same mental process that allows us to politicize the most general—and seemingly universal—of all concerns: the environment.

Democrats have ostensibly become the party of environmental stewardship in the United States, and Republicans the antiheros, despite the fact that, until quite recently, Republicans were known for being the party of conservation and stewardship.

But all of this is a false dichotomy, one that's revealed to be a sham when you consider the fact that President Biden, who emphasized the dire importance of addressing climate change, has approved more oil and gas drilling leases on public land than did former President Trump. At the end of the day, it appears that the leadership of neither major political party cares all that much about the environment.

Meanwhile, even the most natural of allies have been divided and reduced to infighting—a barrage of friendly fire that even Daniel Curry, with his loner tendencies and desire to commune with nature, has been caught in.

Chapter 9

WOLF POLITICS

On August third, 2020, at ten o'clock in the morning, Tim Coleman's phone rings. Coleman, a lanky sixty-five-year-old, answers the call at his home in remote Ferry County, a mountainous area of Washington State near the Canadian border.

On the other end of the line is Donny Martorello, the wolf policy lead for the state at the time. Martorello is bearing bad news: Coleman has been kicked off the state's Wolf Advisory Group, and will be receiving formal notice soon. "I hope we can still be friends," he tells Coleman.

"That was kind of funny," Coleman says afterward, about the offer to be friends. "I'm really kind of floored. It's rage and sadness."

Although he maintains that the news was unexpected, for anyone paying attention to the Wolf Wars, and the broader politics of conservation, the decision to kick Coleman off the advisory group didn't come as a shock. Coleman, who has lived in rural Washington since the mid-1980s, has worked in one way or another on environmental issues for the past forty years. In that time, he's developed a reputation as a blunt and fiery advocate with a varied resume. He's fought logging sales and championed beaver introduction. More recently, the complex issues around wolves have dominated his time, along with the responsibilities of being the Executive Director of the Kettle Range Conservation Group, a non-profit founded in 1976 that seeks to "defend wilderness, protect biodiversity, and restore ecosystems" of the Columbia River Basin in northeast Washington.

The much-newer Wolf Advisory Group, also known as the WAG, was formed in 2013, several years after wolves started returning to Washington. The WAG is a citizen advisory group whose members are appointed by the governor. The stated goal of the committee is to give a diverse group of people a voice when it comes to wolf management in the Evergreen State, and to "recommend strategies for reducing conflicts with wolves outlined in the state's Wolf Conservation and Management Plan." Ranchers, hunters, environmentalists, and conservation leaders all sit on the WAG, and have a hand in crafting policy recommendations, including the Wolf-Livestock Interaction Protocol, which sets the standards for range riding.

WAG committee members trained with the Washington D.C.-based Center for Conservation Peacebuilding on how to work together. To that end, CPeace co-founder Francine Madden spent two hundred days in Washington State and seven thousand hours on the phone. All told, Washington taxpayers paid Madden more than $1.2 million for her guidance.

The entire effort represented the culmination in a decades-long shift in how environmental activism and advocacy has been conducted.

Despite that effort, seven days after the warning from Martorello, Coleman is formally removed from the WAG due to "a pattern of behavior that has eroded necessary trust between yourself, other WAG members, and Department staff," writes Kelly Susewind, the Director of the Washington Department of Fish and Wildlife.

Coleman's is a complicated story, but worth rehashing, because it sheds light on a troubling reality: wolves have divided the most natural of allies, the conservation community.

One month prior to Coleman's removal, I spent several hours with him and his wife, Sue, in the Kettle Range, prime habitat for wolves. We met on the flank of the mountains in thick subalpine trees. A quick, steep bushwack brought us to a trail, and after a mile or so of hiking, the hillside above us opened up, displaying a steep expanse of grass, wildflowers, and weather-bleached deadfall. The Colemans came to a stop on the trail in front of me.

"This is where I saw the wolf," Tim Coleman said, pulling out his cellphone to show me a photo taken years ago. The image was grainy from being zoomed in, but clear enough to make out the details. Coleman recalled the experience with verve, describing how the animal's eyes locked on his, how they shared a moment that seems universal to wolf encounters.

We were making our way toward Scar Mountain, a seven-thousand-foot peak that neither Tim nor Susan Coleman had climbed before. It was a warm day in July, but both of them moved fast. Veteran mountain climbers and life-long activists, they were keeping an eye out for wolf sign.

We found plenty of it: at least five large and relatively fresh piles of scat. This made sense—the area we were hiking in had seen some of the region's most heated wolf-livestock conflicts. Since 2012, the Washington Department of Fish and Wildlife has killed twenty-eight wolves believed to have killed cattle in these hills. Most of the cattle attacked or killed have been owned by the same ranching family, the McIrvins.

We reached the summit of Scar Mountain and found a rocky bench on which to have lunch. From there we could see southeast into a wide valley. That valley was where the McIrvins, the owners of Washington's most contentious and controversial cattle operation—the Diamond M Ranch—grazed most of their cattle, Tim Coleman explained to me.

Since 1985, the Colemans have lived off-grid, relying on solar power and water pumped from a spring. They grow much of their own food, raise chickens, and brew their own beer. Sue Coleman manages the local health food store.

Tim Coleman started climbing mountains, and paying attention to the environment, while he was stationed in Japan serving in the US Navy. "You'd get to the top of a mountain in Japan and there would just be a big pile of cans and trash," recalls his friend and fellow environmentalist, Mike Petersen.

After Coleman and Petersen discharged from the navy, they lived in Portland for a time, which is where Tim met Sue. Eventually, the three of them bought property in northeastern Washington and immersed themselves in the environmental issues of the day, namely logging.

During the 1980s and '90s, the Timber Wars were passionate, sometimes violent affairs. The poster child of these conflicts was the northern spotted owl, whose fate depended on the old-growth forests that logging companies sought to cut down. Inspired by Edward Abbey's novel *The Monkey Wrench Gang*, activists took to the forests, chaining themselves to trees, sneaking onto timberland, where they occasionally sabotaged logging equipment. The animosity went both ways, with Forest Service personnel driving unmarked cars to avoid the ire of timber towns.

Coleman and Petersen were active participants in anti-logging protests. They attended Earth First! events and did their own bit of direct action. Petersen, who was the director of a regional conservation organization until he himself was ousted for political reasons, demurs when asked how far over the line he went with his activism.

The Timber Wars culminated with increased protections for the spotted owl under the Endangered Species Act. As the '90s drew to a close, the logging of old-growth forest in the West slowed dramatically, and the erstwhile firebrand environmentalists found themselves with some political and social clout. They finally had "a seat at the table," according to Petersen. And so their tactics evolved. Going forward, the name of the game was collaboration. In fact, both Coleman and Petersen now collaborate with a regional logging company, a state of existence unimaginable during the heat of things three decades ago.

This model of conservation work—prioritizing collaboration over confrontation—has infused most contemporary environmentalist efforts. Although there are still organizations embracing radical tactics, far more groups are trying to form bridges than spike trees nowadays.

The notion of collaboration forms the ideological basis for the Washington State wolf plan, as well as the WAG. And Coleman isn't opposed to collaboration, per se.

But as he witnessed the wolf debate seemingly spiraling out of control, Coleman started to wonder if the collaborative approach was the right one, and over time he has become increasingly outspoken, going so far as to get involved in several lawsuits aimed at limiting the killing of wolves. And this, according to the WDFW, was a bridge too far.

Coleman's formal dismissal from the WAG cites three distinct events that broke trust.

First, in 2017, he provided a declaration in a court case against the state, challenging the killing of some wolves. Second, in 2018, he appeared in a highly critical documentary called *The Profanity Peak Pack: Set Up and Sold Out*, which was produced by Predator Defense, a group that eschews the collaborative model of conservation. Coleman's actions ruffled enough feathers on the WAG that other members brought up their concerns about him during a 2019 meeting. This resulted in a change to committee bylaws in April 2019, requiring all WAG members to provide "advance notice to their fellow WAG members and the Department when they were taking action that was not in support of WAG-based decisions, such as litigation, news releases, videos, etc., so all would have full knowledge ahead of time and not be caught off-guard."

Which brings us to the third infraction. Late in the summer of 2019, Coleman became a litigant in another wolf-related court case against the state, this time challenging the policy allowing the state to kill wolves. The final sin Coleman committed, according to the department, was breaking this newly agreed-upon tenet to disclose his involvement with the lawsuit ahead of time.

And this is where it gets even messier. Coleman maintains that he did send advance notice. However, due to technological issues, he didn't know that the email he'd sent hadn't gone through until the court case had been filed.

"I apologize for this error and among the many complications it caused most importantly including advance notice as per our WAG agreement to notify you all of a decision I had made to sign on as plaintiff in a lawsuit regarding the OPT Pack removal," he writes in an August twenty-ninth email entitled "Mea Culpa."

WAG member Tom Davis responds, thanking Coleman for the explanation, but calling into question his commitment to the collaborative process. "From my vantage point it appears that you are trying to walk the path of conflict transformation while at the same time maintaining the role of an activist who criticizes the department's management of wolves publicly and partakes in lawsuits against the department's wolf management," Davis writes, cutting neatly to the core of the issue. Davis, the director of government relations for the Washington State Farm Bureau, goes on to write, "I can only speak for myself, but I do not think that is helpful to what we are trying to accomplish as WAG members."

Coleman believes there are other reasons behind his dismissal.

Earlier in 2020, he signed onto another lawsuit. This one was against the Colville National Forest, which had just wrapped up a multi-decade overhaul of its forest management plan.

This time Coleman did notify fellow WAG members, although in an email to them he pointed out that the lawsuit "does not challenge WDFW wildlife management nor does it in any way implicate the WAG or State of Washington."

Other members didn't see it that way. Davis, the Farm Bureau director, responded quickly, calling the WAG process the "finest example of collaboration" he's ever been involved in and alleging that Coleman's newest lawsuit could "cause financial ruin to our rural communities."

"In your past legal actions Tim I have been tempted to call for your removal from the WAG, but I refrained. I can no longer remain silent," he wrote. "Therefore, I formally request that Director Susewind remove you from the WAG."

Ten weeks later, Susewind does just that, informing Coleman that "your participation as a litigant against the US Forest Service in 2020 is not the basis for your removal, but it undermines the ability of the group to see you as a true collaborator and brings up past issues in which you did violate WAG commitments."

The story ripples throughout the conservation community. Anti-collaborators point to the Coleman dismissal as more proof that reaching across the aisle only leads to harm. At the next WAG meeting, impassioned public commenters decry the removal. Steph Taylor, an employee at an animal rights group in Washington, articulates the sentiment many are feeling following the news of Coleman's removal.

"The WAG has removed the only true conservation organization on the WAG," she says.

"You will be forced to betray your principals," another speaker warns the WAG's remaining conservation groups.

Coleman agrees with this assessment, calling the WAG process a "ruse."

"This is not a process that welcomes diverse opinions," he says. "It's a process that is very carefully engineered to get the right people on the WAG to go along with what the department wants."

Notably silent from the public outcry are the three conservation groups left on the WAG: the Humane Society of the United States, Wolf Haven International, and Conservation Northwest. "My caucus," Coleman had called them prior to his removal.

From the outside, this division amongst the conservation community is confounding and hard to understand. Why would seemingly natural allies not have Coleman's back? In many ways, it comes down to a single question: Should wolves be killed, ever?

The bottom line is that these other conservation groups on the WAG have come to the conclusion that killing wolves must be an option for wildlife management agencies. This doesn't mean they like it; they have asked for rules governing when and how the killings happen. But their flexibility on this issue differentiates them from the groups and organizations that howl whenever—and however—wolves are killed.

And then there are the conservationists who are coy about their true beliefs. Some organizations and individuals publicly won't support the killing of wolves, but in private will admit that it must be done on occasion. Others will privately criticize the killing of wolves by the state or federal government, but, for one reason or another, remain tight-lipped about raising their concerns in public.

As in all politics, those with money have the power. After attending the swanky dinner in Seattle, Daniel Curry finds himself with a new, powerful (albeit hesitant) ally. Jennifer McCausland, the wealthy

Australian émigré, begins supporting the range rider's work, even if she remains skeptical of range riding writ large. In fact, she gives him five thousand dollars to pay for hay during the winter.

However, McCausland is also a vociferous critic of Conservation Northwest, a nonprofit that gives Curry money for his work. She reserves particular disapproval for Mitch Friedman, the executive director of CNW. She says she's "always disagreed" with how Friedman handles wolf issues. "He's a danger to the wolves. He's a danger to all the animals because he's willing to forfeit their lives for a quid pro quo."

The "quid pro quo," of course, is Friedman's willingness to work alongside ranchers and others who don't hold the same ideological beliefs as he does. Nevertheless, McCausland decides to help pay for some of Curry's hay. To do so, she donates the money to the Lands Council, a smaller conservation group based in eastern Washington that at that time took a harder line toward wolf recovery than does CNW—they do not support killing wolves. The Lands Council then gives the money to Curry.

Meanwhile, in a ridiculously petty twist, McCausland's stepdaughter donates one hundred dollars to Conservation Northwest in her mother's name as a birthday present.

McCausland is furious. She emails Mitch Friedman and tells him that in no way, shape, or form does the donation represent her endorsement of CNW's approach to managing wolves. She goes on to inform Friedman that she's decided to support Daniel Curry through the Lands Council.

Friedman, in turn, gets angry, worried that the Lands Council is trying to poach his star range rider. All of which leaves Curry in a perilous spot. He has already expressed concern over the lack of control he has over his work. Finding himself caught up in the complex politicking of the conservation community—the same community he relies upon for his annual funding—is a dangerous spot to be in.

"She will give you money," one veteran conservation organizer tells me about Jennifer McCausland. "A lot of it. But then she will try to direct what you're doing."

This is hardly unique to McCausland; few organizations are willing to sign a check without having some oversight over how that money is used. Curry needs money to do the work that he believes in. But to get funding, he has to bow to the will and desires of those with money.

"The real problem isn't wolves or cows," he says. "I've known that for a long time. It's so polarizing. It's so convoluted. If it's just wolves or cows, no problem."

Daniel Curry has had his own brushes with controversy. In early 2019, after I'd written about allegations of fraud against other range riders in eastern Washington, and several months after I'd published a long profile about Curry in the newspaper that employs me, I received an anonymous email.

The fraud story was based off an investigation conducted by WDFW law enforcement. They found that in 2018, two state-contracted range riders who were supposedly protecting cattle in Ferry County had actually been more than one hundred miles away, shopping and spending time in an upscale hotel. While those riders were AWOL, wolves ended up killing some cattle, which led to the extermination of the wolves.

All told, the range riders were accused of stealing about $2,000. The state requested second-degree theft charges. For their part, the riders denied the allegations. A few days after the story went to print, I received the following email:

> Range rider Daniel Curry has a case of fraud against him by the WDFW buried somewhere on the prosecutors (*sic*) desk. It is the result of an investigation involving him and Jay Shepherd from as far back as 2014, I believe. It is a much bigger case involving a lot more money and people. Fact!! It is public record at the WDFW but nobody knows about it . . . yet. At some point it will come to light but not before the man hustles more money from unsuspecting, honest, hardworking people like yourself, via Gofundme (*sic*). Just sayin.

I knew that WDFW had fired Curry—or rather, in sanitized *bureauspeak*, that his contract had not been renewed. He had told me that this was because he'd used department money to pay a range rider who had never worked for him. However, he was quick to point out that he'd been directed by his supervisor Jay Shepherd to do so, and had documented the transaction meticulously. Aside from this episode, the allegations in the anonymous email were unfamiliar to me.

In a subsequent email, the sender said that they didn't want to tell me more, for fear of revealing who they were. The allegations disturbed me, especially because I'd already started work on this book. I told Curry about them, and he asked me to show him the email so he could try and figure out who sent it. I declined, suddenly questioning his trustworthiness, then filed a passel of records requests with the Fish and Wildlife Department and waited anxiously.

Several months passed, and the records from the state started to trickle in, each of them tantalizingly incomplete. Nevertheless, the picture began to clarify.

Range riders in Washington work via a contract system. Starting in June 2015, and lasting until June 2017, Curry's company GRIPH was under contract with WDFW. Over the course of those two years, Curry invoiced WDFW for $204,210. That money went primarily toward horse feed, gas, and, crucially, the wages of other range riders that Curry hired.

In 2017, he reapplied for the contract through a competitive bidding process. In June of that year, he learned that his bid had been denied after WDFW officials determined him to be, in their estimation, "not a responsible bidder." An internal audit conducted by the agency revealed that Curry had invoiced the state for one range rider who didn't work for him. Additionally, the agency claimed that Curry didn't have proof of payment for an additional $12,600. According to the letter sent to Curry, the case had been referred to the state auditor's office.

Daniel Curry was a fraud, or so they said.

Range riding is primed for fraud. Riders spend days out in the field, alone or with just one other person. They are often out of cell range, far from towns, roads, and administrators. On top of that, the necessary skill set is diverse and hard to manage.

Carter Niemeyer has worked with wolves for more than thirty years, first as a government wolf trapper, then as the federal wolf manager in Idaho, and more recently as an advocate and educator. If anyone knows the possibilities and perils of range riding, Niemeyer does. Range riders need to know wolf and cow ecology, he tells me. They need to be able to ride horses, cut trail, and live in the wild. They need to be hard workers. In Niemeyer's opinion, most people can't or won't do that.

"There is an expertise there," he tells me, "and most people who engage in range riding are not experts, and need to be schooled and trained by those who do it professionally. Where it gets off to a bad start is there is a pot of money . . . people come running and say, 'How do I get some of it?'"

This is what appeared to be happening in the WDFW range riding program. Unscrupulous riders were fleecing the public and, incidentally, leading to the deaths of wolves. But was Daniel Curry one of these scoundrels?

The records about Daniel Curry trickled in, but the final delivery date kept getting pushed back. First into the summer, and then—due to a new COVID-19 exemption—late into the fall. I knew what he was accused of, but I didn't know whether the accusations were substantiated. He maintained that the department was out to get him, reason unknown. He told me that he'd sent the required documentation, and that he'd been directed by his supervisor at the time to pay the range rider who didn't work for him.

I found both of these points hard to believe—first, that the department would have any sort of vendetta against him, and second, that he had been ordered to commit fraud. And yet, unlike with the other range rider fraud case, the department wasn't recommending that theft charges be brought against Curry.

Hoping to speed up the process, I contacted the state auditor's office and left a message explaining what I was looking for. I expected the slow and Sisyphean process to drag on, and resigned myself to the wait.

They called me back that same afternoon.

"We didn't find anything significant," said Kathleen Cooper, the spokesperson for the state auditor. In fact, Cooper told me, the only reason WDFW was investigating Daniel Curry was because the state auditor was investigating WDFW's range rider program as whole. During a routine audit, they'd pulled a number of range rider contracts, including Curry's, and asked for documentation. The state didn't have it. In a December 2018 letter to WDFW, the auditor explained that the issue was the agency's responsibility, rather than Curry's. And it turned out that the auditor's office had already pointed this out to WDFW during a previous audit. "In our fiscal year 2016 accountability audit report, we recommended the Department improve internal controls over monitoring of Range Rider contracts to ensure adequate oversight and monitoring to safeguard public resources," the letter stated, a rather dry admonishment. Another employee I spoke to at the auditor's office put it more bluntly: "The department was trying to cover their butt." The bottom line was that the allegations in the initial email which had concerned me so much weren't true. Just another layer of infighting in the wolf world.

However, for Curry, the damage was already done. He didn't get another contract with WDFW. Now, like many ranchers and others who live in rural Washington, he distrusts the wildlife department with a sizzling resentment.

That distrust of the WDFW extends beyond ranchers and spurned range riders. Environmental groups—at least the ones on the *never kill* side of the wolf wars—also question the agency's every move. Not long after he lost the contract, Curry began to range ride for the Northeast Washington Wolf-Cattle Collaborative, or NEWWCC.

Founded in 2018, the organization's goal was to fuse the sort of theoretical knowledge utilized by the larger state agency with the embodied and lived experience of ranchers and farmers. NEWWCC's founders, Jay Shepherd and Arron Scotten, hailed from these two disparate worlds.

Shepherd, who worked for the Washington Department of Fish and Wildlife as a biologist, and grew up on a farm and ranch in eastern Washington, also has a PhD in Natural Resources. He's an employee of Conservation Northwest, the organization that Jennifer McCausland loves to hate.

Meanwhile, Arron Scotten claimed to be a fifth-generation rancher and military veteran with strong local ties. Their collaboration was heralded as a possible way to promote nonlethal deterrents for wolves, thus saving ranchers the loss of their livestock and saving wolves their lives.

"Wolves add a new cost to cattle operations that are part of the culture up here," Shepherd told the *Spokesman-Review* in 2018. "We want to provide more access to nonlethal depredation deterrents."

Shepherd's and Scotten's connections to the area seemed likely to assuage one of the most common complaints aimed by locals at pro-wolf groups—that these preachers of coexistence have no idea what it's like to live near predators.

But it didn't take long before rumors began to spread: Arron Scotten wasn't actually a fifth-generation cattle rancher, not by any stretch of the imagination. Nor, some claimed, was he a veteran. These unsubstantiated rumors began to swirl. And then WDFW announced that they were recommending that second-degree theft charges be brought against Scotten and his wife Jolene, as well their company, DS Ranch. The agency alleged that the Scottens had been in Spokane when they were supposed to be in Ferry County, on duty as range riders.

When I connect with Jay Shepherd, he tells me that his former partner had wanted to be paid by the organization. After learning that it was illegal for a board member of a nonprofit to be paid by the nonprofit they serve, Arron Scotten decided to resign from his seat on the board of NEWWCC on November ninth, 2018.

Shortly after Scotten resigned, Shepherd and others on the board of NEWWCC became aware of the WDFW investigation and elected not to hire him. Shepherd maintains that he knew little about the investigation other than that it was happening. He does believe that Scotten's decision to bring his family members onto the WDFW contract was a mistake.

"There was a hell of a lot of money at stake," says Shepherd. "Arron's family kind of became the face of WDFW in Ferry County. And there was some bitterness in all of that."

WDFW detective Lenny Hahn started investigating the DS Ranch on October fifteenth, 2018. At the time, the range riding associated with the business was valued at $352,000, and listed seven different people, including Jolene and Arron Scotten, several of their relatives, and Jay Shepherd.

According to cellphone records obtained via search warrant, there were numerous times Scotten and his wife claimed that they were working when they were not. Some of those periods aligned with times when wolves attacked cattle. And this is where the Scottens' deception cost the state more than just money, says Chris Bachman, the former wolf program manager at the Lands Council. (Bachman now works with Tim Coleman.)

Washington's lethal removal policy allows for the killing of wolves if they kill or injure livestock three times in a thirty-day period or four times in a ten-month period—but only if two nonlethal deterrents have already been deployed.

On September twelfth, 2018, WDFW announced the planned killing of members of the Old Profanity Territory (OPT) wolf pack following repeated attacks by the pack on cattle. According to a news release, the Diamond M Ranch had used several nonlethal deterrents to try to fend off the wolves. These efforts included range riding, calving outside of the wolf pack's range, delaying the turnout of calves until they were larger (and harder to kill), removing livestock carcasses, and removing sick and injured livestock.

According to WDFW documents, the justification to kill some OPT wolves (eventually the agency would kill all four known members of the pack) was made in part because the agency believed there had been heavy range rider presence in the area.

Meanwhile, the investigation alleged that Scotten and his wife, who rode primarily in the OPT pack area for the Diamond M Ranch, were not working when they said they were. For instance, Scotten claimed to have worked a combined twenty-five hours on September fourth and fifth, 2018. In reality he was in Spokane buying building supplies, the WDFW investigation reported. During that same two-day stretch, the OPT pack injured two calves and killed one calf.

A few days later, on September twelfth, WDFW Director Kelly Susewind authorized the killing of members of the OPT pack.

Six days after Susewind's approval for lethal removal, Scotten claimed to have worked seven hours. However, according to the investigation, he was staying at the Davenport Hotel in Spokane on September eighteenth and nineteenth. Jolene Scotten also claimed to have worked eight hours on September nineteenth. The investigation revealed that phone records also placed her in Spokane.

On September twenty-first, the OPT pack attacked and injured five calves. Meanwhile, some at WDFW were questioning the range riding efforts in the area.

"Staff had not seen the range riders during most field checks or on most trail camera photos," a Ferry County WDFW staff member informed the agency's Eastern Region director on October twenty-fifth, 2018. Nevertheless, the following day, WDFW authorized another round of lethal removal.

And yet, out of 440 fifteen-second videos taken by WDFW trail cameras in the OPT pack area throughout September and October, riders with DS Ranch were spotted only four times.

The concerns persisted. In notes from a call in July 2019 discussing the OPT pack, employees in the region questioned whether "actual range riding, not just driving on the road" had ever taken place. "Have never had actual, quality range riding on this landscape," the notes went on. "Daily patrols aren't doing much."

These documents were all provided to me by the "never kill" caucus, and, in some ways have ended up even further fanning the flames on the conflagration between the various conservation groups who have a stake in the management of wolves in the Evergreen State.

The "never kill" caucus points to this latest controversy as yet another example of a conservation group having sold its soul. And it's true, the optics aren't great: even though NEWWCC is not directly run by Conservation Northwest, it does receive some of its funding from them, and Jay Shepherd does happen to be a CNW employee.

"I want to see range riding work," says Bachman, the former Lands Council program manager. "I want to see it done well and right. This kind of stuff is what is in the way of that happening."

Joel Kretz is a rancher-politician with a penchant for theatrics and dislike of predators (cougars and wolves, in particular). For more than a decade he's advocated for ranchers in working to ease the transition from a wolf-free landscape to a wolved one. In that time he's posed for a number of portraits at his horse and cattle ranch. Something about the man—perhaps his television-ready cowboy aesthetic—drives otherwise sober photojournalists to go to extremes in their attempts to create the perfect Western portrait.

In one photo published in the *New York Times Magazine*, Kretz sits on a bale of hay, wearing a white Stetson, his thick mustache sharply bifurcating his face. A dog sits on his lap, another stands behind him, and the flanks of grazing cattle are visible in the background. The sky is dark and moody and Kretz looks dour, a perfect picture of steely determination. In another newspaper photo he leans on a log fence, late afternoon sunlight bathing his face as he stares into the distance while cradling a .30-06 hunting rifle.

He was elected to the Washington State Legislature in 2004, and his advocacy for the rural way of life has gained him a devoted base of constituents. On social media he depicts himself as a man reluctantly forced into the halls of political power. A lone voice of conscience fighting the good fight.

For a man who was born on Mercer Island, deep in the heart of liberal Washington, this has been a skillful political trick to pull off. As a state representative he's worked on many issues, even gaining some fame in the beaver advocacy world for helping craft some of the most beaver-friendly legislation in the nation. However, whether he likes it or not, Kretz is best known for his work with wolves. And, while he often speaks disdainfully of the machinations of politics, he's not above getting dirty.

In 2017, Kretz helped cut funding for a controversial Washington State University wolf researcher, Robert Wielgus, who'd published research showing that killing wolves didn't reduce wolf attacks on cattle. Wielgus, an outspoken and flashy character himself, had also accused the Diamond M Ranch of intentionally putting their cattle near wolves.

Infuriated by Wielgus's research and allegations, Kretz and other lawmakers eliminated the funding for the scientist's research.

Kretz likely knew he held some leverage over the university; at the time, WSU was in the midst of securing state funding for a medical school. WSU officials worried that the controversy threatened funding for that project, and in 2018 Wielgus left WSU (but not before winning a three-hundred-thousand-dollar settlement from the university).

All this politicking drives Kim Thorburn a bit mad. Thorburn is one of nine governor-appointed commissioners overseeing the Washington Department of Fish and Wildlife. Among other responsibilities, the commissioners vote on changes to hunting regulations and approve land acquisitions.

Thorburn is a no-nonsense woman who, prior to serving on the commission, served as the public health officer for Spokane. She's a scientist and a bird lover. She was forced out of her job as a health officer for unspecified reasons, although many believed it was because she was too blunt.

She's found herself inhabiting a similar role on the commission, regularly articulating points of view that others hesitate to express. For example, Thorburn believes that wolves dominate the conversation more than they should. She's resentful of the resources that state and federal government pour into managing wolves, while other species that are teetering close to the brink of oblivion receive scant attention.

Sage grouse are one such overlooked species. The ground-loving bird is perilously close to extinction, with much of its habitat chewed up by agriculture or burned up during increasingly severe fire seasons. Yet the state employs only one fulltime sage grouse biologist, despite having once had a widespread sage grouse population.

Meanwhile nine WDFW employees are totally and solely focused on wolf recovery.

"Wolves suck up these resources," says Thorburn.

Beyond diverting scarce and much-needed resources, the politicization of wolves—and, by extension, wildlife management—has another grave implication: If the fickle winds of politics dictate how wildlife management decisions are made, how can professional wildlife managers be expected to make sane and sound decisions?

In a large conference room flanked by the mounted heads of deer, elk, bears, and at least one stuffed cougar, the tension between biology and politics is on full display in June 2021. The Idaho Fish and Game Commission has convened to undertake the task of amending the state's wolf hunting and trapping seasons in response to a newly passed—and highly publicized—law aimed at drastically reducing the wolf population in the Gem State. During the hour-long conference call, the seven commissioners express discomfort with the duty at hand. Namely, making biological decisions via legislative fiat.

"I think this could have been handled so much better," says commissioner Don Ebert. "I wish the legislature would be partners with us."

The critique is notable because Ebert broadly supports the liberalized wolf-hunting seasons.

Senate Bill 1211 established a year-round trapping season for wolves on private property. The bill also allowed the unlimited purchasing of wolf tags, and authorized wolves to be taken by any of the approved methods for trapping other wild canines (foxes, coyotes) in Idaho.

The bill has garnered national and international attention, particularly one provision, which calls for a 90 percent reduction in Idaho's wolf population, currently estimated to be about fifteen hundred. The law went into effect in July 2021.

Some of the commissioners—all of whom were appointed by the governor and confirmed by the state senate—have described their mandate as a needle-threading exercise. In addition to amending the hunting and trapping seasons, the commission is responsible for rewriting the current hunting rules and regulations so that they will align with the new legislation.

Brad Corkill, the chairman of the Fish and Game Commission at the time the new legislation was passed, says that he was notified about the impending law less than twenty-four hours before it went to vote.

"I find that a tad bit disrespectful and insulting on part of the legislature," he says during the conference call. "They dumped this in our lap . . . giving us very little options as to how to handle this situation. Disrespectful is the kindest word I can come up with on this."

Numerous conservation and environmental groups have decried Idaho's new wolf law. In late May 2021, the Center for Biological Diversity, the Humane Society of the United States, and the Sierra Club jointly filed an emergency petition asking the federal government to relist wolves in the Northern Rockies as an endangered or threatened species. The US Fish and Wildlife Service must respond to the relisting petition by August twenty-fourth, and could potentially wrest the management of Idaho's wolves away from the state.

However, wildlife managers for the Idaho Department of Fish and Game are skeptical that the new law will have a significant impact on the overall wolf population.

"At the end of the day, wolves are part of the landscape, and I don't think you're going to see that change," says Chip Corsi, the agency's regional manager in Coeur d'Alene. "We've managed them pretty aggressively basically out of the gate. I think the guys who are hardcore wolf trappers will tell you it's not easy to trap wolves."

The new legislation in Idaho is just the latest example of biological and ecological decisions being made through the political process. Recently, a ballot initiative in Colorado narrowly approved the reintroduction of wolves into that state, despite wildlife officials' warnings against doing so. And Montana and Wyoming have joined Idaho in introducing and passing a number of laws specifically targeting wolves.

"These kinds of legislative means can cut both ways, depending on who is in power," says Dave Ausband, a professor at the University of Idaho who studies wolves.

Whether or not Senate Bill 1211 and the subsequent hunting and trapping rule changes have a meaningful impact on Idaho's wolf population "remains to be seen," says Ausband. "Even if you mailed every Idaho citizen a wolf tag, it doesn't mean 1.7 million people are going to be out there hunting wolves."

More troubling to him is the mixing of politics and biology. The North American Model of Wildlife Conservation (the set of principles that guides wildlife management and conservation in the United States and Canada) largely relies on fishing and hunting license sales to fund state wildlife agencies, intentionally separating politics and biology. That model grew out of the excesses of market hunting and other practices which nearly led to the extinction of now-common species like elk and deer.

Under the model, state fish and wildlife agencies receive the majority of their funding from hunters and anglers, giving them some separation from the machinations of politicians.

"There is a reason there is a distance between elected officials and state wildlife agencies," Ausband says. "The [Idaho] legislature . . . completely closed that distance. That gap. To me that's problematic. That's not our tradition. That's not our legacy.

"If you don't like wolves right now you might support that decision. But what if your guy or gal isn't in control anymore and it's the opposite? Is that really the way you want to do things? This crazy pendulum that rockets back and forth?"

Chapter 10

ANOTHER WAY

On a cool evening in late May, politics and pendulums are far from sight. It's been a long day, with ten of us laboring away, but now the drive of work has subsided and we're relaxing in Jerry Francis's shop, looking out over a snow-fed creek and greened-up pastures. All day we've been rounding up his two hundred or so cattle, separating the cows from the calves, getting them vaccinated, castrating the calves, and tagging them. Francis, lean and a little stooped in the shoulders, is wearing worn jeans and a tattered Washington State Cougars T-shirt. He doesn't talk a lot, but when he does, the assembled folks listen.

If you live in the western United States—or anywhere where American movies are common fare—you already have a basic mental image of what a cattle roundup looks like. Cowboys on horses. Cattle bawling. Lots of coffee. Dust. Lanky men yelling and cracking whips.

On Francis's ranch, there's coffee and dust and human bodies made lean from hard outdoor work, but the cattle rarely bawl, and everything that's done is done intentionally and slowly with cow psychology held firmly in mind.

About a decade ago, Francis watched a film about low-stress cattle management as championed by Mary Temple Grandin.

Her message stuck with him and prompted him to change his approach.

Francis lives on his well-kept ranch, which is nestled in a lush valley bottom just spitting distance from the Canadian border. He is a multigenerational rancher, but unlike his father, this work is a side gig for him; he holds down a day job selling tools to pay the bills. It's evident that he is also one of those rare human beings who, upon taking in some new information, manages to buck tradition and habit and actually change.

The Grandin film made a lot of since to Francis. The basic premise being: if we consider how cattle think, and design ranches to meet their needs, the animals will be happier, and handling them will be easier. Cattle see out of the sides of their heads, whereas humans and other predators have eyes that look straight ahead. This difference is fundamental to cow psychology. So Francis changed his ways and started to move and handle his cattle in a slower and calmer manner. "Less of that yippee-kay-yay bullshit," as one of his ranch hands says.

That morning, Daniel Curry and I arrived at the ranch at 9:00 a.m. We spent the first hour or so rearranging the cattle gates. While building chutes and pathways, we imagined how the animals would interact with the fences. Was this too tight of a corner for a five-hundred-pound cow? And what about this turn-around point, or that dark corner?

Once that was done, Mike, Francis's right-hand man, jumped in an old beat-up Ford with a large bale of hay strapped to the flatbed and drove into the pasture where the cattle were lazily grazing.

"You guys go stand over there," Francis said, gesturing to a cottonwood tree near the creek. Mike drove slowly, using the tasty hay to entice the cattle to move toward the fences we'd just constructed. We stayed off to the side, out of sight near a creek.

The cattle fell in behind, their *moos* echoing off the nearby hills. Leaning against a cottonwood, Daniel Curry pointed up the hillside where the field turned to trees and the slope steepened. "Right there," he told me, "I saw a pack of wolves." Curry had spent a bitter winter week here hidden in the top of an old barn like some World War II sniper, guarding against wolves. Whenever they did appear, he'd flood them with light and fire noisemaking rounds from his gun. Eventually the pack got the message and moved on.

"Jerry's got a little Serengeti here," he said, listing off all the moose, elk, deer, and other wildlife the rancher has seen strolling through his backyard. "The number of wolves he could have killed—"

Curry trailed off as the cattle picked up the pace chasing the truck. Some broke off, heading in the direction of the creek, so we hiked out from the cover of the cottonwoods and shooed them back to the main herd. "When moving cattle," Curry reminded me, "imagine how they see—or don't see—you. Don't go at them straight ahead. Come in at an angle so they know where you're at."

Months earlier I learned this lesson the hard way when wrangling cattle with Curry. We were riding Griph and Raven while pushing a herd through a pasture in northeast Washington. I approached a group of cows too directly and they bellowed and scattered. It took us more than an hour to regroup them, a job that would have taken minutes if I'd properly hacked their cow brains.

Back on the ranch, the stragglers rejoined the main herd, and we funneled them into a penned enclosure, separating calves from cows. Here the bawling increased, but Francis, ever attentive to cow minds, had designed the pens in such a way that the calves and cows could touch noses, keeping in close contact and easing some of the stress of separation.

Francis handed me a pair of worn work gloves and placed me in front of a devilish-looking pinching machine. "Don't hesitate," he told me, then headed off to supervise another part of the operation.

The day's work really began. Cows and calves entered a tight chute, pushed forward by the pressure of their brethren behind them. The only way to exit this chute was to step out and onto the device I was operating, a metal pincher that squeezed their bodies, immobilizing them.

Once immobilized—and no matter how gentle or low-stress I might try to be with them, they didn't appreciate this—a more experience cowhand would get to work while I held the metal lever tight, keeping the plates pressed against the struggling body. For the cows, they got a quick round of vaccines. One jab, two jabs, and then a quick examination of ear tags and an overall pat down. When released, they roared out of the chute, spring-loaded with a new sense of freedom, and raced into another pen where they licked their wounds.

Calves, on the other hand, were a different story. My job was similar: slam shut a metal pincher that caught them around the outside of their neck. Once trapped, I closed the rest of the metal contraption, locking their bodies in place.

Then Curry gave them a shot, and the cowboy grabbed them by the ear, and, using a large device that looked a bit like a hole punch, slammed tags into their ears. This procedure elicited the biggest bawling, for good reason. The holes were huge.

Finally, another cowboy checked to see if each calf was a male or not. The boys were going to lose their balls; the cowboy reached under and clipped a rubber band around them. Over the next few days, the testicles would fall off, offering a tasty snack to some wild animal.

"Those poor guys will sulk," said Mike. "They know what they lost. And they fight like hell in the moment."

Francis knows the cows well, including which calf belonged to which mother. "That's a stubborn one," he said, pointing at a truculent youngster. "Always has been, ever since he was born."

More than one calf squeezed out, my clamping neither fast nor firm enough. As a city boy, firmness doesn't come naturally to me. I struggled, particularly with the calves, hesitating to apply so much force. But I learned quickly that if I didn't slam the gate closed fast enough, I would end up hurting them more. It was better, then, to be firm and decisive.

The consequences of not doing so were illustrated by one black calf, which despite being a youngster still weighed upward of two hundred pounds. Stubbornly he backed into the pen. "That's fine, we'll make do," Curry told me. So I yanked on the lever, squeezing his body tight. With his head free, however, he freaked out, lashing around, bucking like the bull he would never become. Before I could react, his head got jammed between the parallel metal bars of the chute. The calf was stuck and jerking. I was afraid he'd break his neck.

Francis, who up until that moment had been standing off to the side overseeing the operation, leapt forward, placing his hands over the calf's eyes to calm him down. Calmly but firmly, he directed us to pull this way and that, manipulating the terrified animal's head until it popped free.

Here is a trait I've noticed among people who spend their lives working with and around animals: a decisiveness and clarity of movement. Perhaps this behavior is learned from the animals themselves.

Daniel Curry once gave me some advice that illustrated the heart of the matter succinctly. "How can you tell the difference between dog tracks and wolf tracks?" he asked rhetorically.

Well, certainly the size is one indication. And shape. But the single best way to tell them apart? Dogs mill, a confusion of paw prints as they sniff every rock and explore every nook. Wolves don't mill. Three wolves, walking in a single line, can cover miles of snow-covered ground without ever stepping outside of the line of the wolf ahead. Oftentimes, trackers will believe they're following one wolf, only to realize a mile or two later that the tracks were left by an entire pack moving in precision. A life exposed to the hard constrains of nature doesn't allow indecision.

For almost a decade, Curry has worked with Francis to help keep wolves from the rancher's herd. In return, the range rider has learned a thing or two about handling cattle—experience which has informed how he works with other ranchers. Watching them in action, it's obvious that the two men work well together. Francis listens to Curry's ideas and implements them when they make sense. Curry, for his part, respects Francis's desire to coexist with wolves.

———

"He's a good guy," Curry says about his colleague. High praise from a man with exacting standards, particularly when it comes to animals.

This is a success story, particularly for the state of Washington, where the return of wolves has included both cultural and biological setbacks—setbacks that can feel cataclysmic for the people whose lives are organized to whatever degree around wolves and their prey. And yet the wolf population in Washington continues to grow: most recently a 24 percent leap between 2020 and 2021 after double-digit growth in previous years.

"I emphatically believe that Washington is the best state to live in if you are a wolf," says Mitch Friedman, the executive director of Conservation Northwest. "Our human-caused mortality rates are far lower. There is much more effort here in trying to minimize conflict and polarization, and I think that's paid off for communities, for ranchers, for wolves. For everybody. Yet the common perception is that Washington is the worst place to be a wolf, and that's because every wolf killed here gets a headline."

Curry concurs with Friedman's general points, but he's still concerned about the outlook for wolves in Washington. He worries about how divisive an issue they have become for people all across the state. He worries about the ranchers who lose cattle. He worries about the individual wolves that are killed by the state or by less-than-scrupulous ranchers. And he worries that the work he is doing, which includes helping a rancher vaccinate his herd on a spring day with no expectation of payment, isn't appreciated. At its best, range riding is more a way of life than a job description. Every high-profile failure casts doubt on the slow and nonlinear way that Daniel Curry goes about his work.

For government agencies and nonprofits, range riding can look like a risky gamble: Does this even work? And if it *does* work on occasion, is it scalable? Can it be implemented throughout the West and not just in isolated triumphs?

"This place has the chance to work out," Curry says about the current state of affairs for wolves in Washington. "It could be a good example. Or it could all fall apart."

I've learned that he often overstates things, and that he has a flair for the dramatic, particularly when it comes to things he cares about. And so I tend to agree with Friedman. Washington is probably the best place to be a wolf in modern America.

But when you zoom in and consider the hyperlocal details of the particular success story that's taking place right here on Jerry Francis's ranch, it's clear that Curry's concerns are valid. How can something as intuitive and relationship-oriented as a range rider helping a rancher round up his calves be scaled up to a landscape level? Or a policy level?

When range riding works, why does it work? And how can those successes be replicated in a way that works for the ranchers, the wolves, and their advocates? These are the core questions of a study started in the spring of 2022 that includes more than six hundred ranchers, two tribes, seven states, and land and policy wonks.

"The study will examine several nonlethal methods of dealing with predators," says Rae Nickerson, a PhD student at Utah State and a member of the research team. "We don't know that much about range riding and what is effective. And people have really different ideas about how to do it."

After being postponed for two years due to the pandemic, fieldwork finally got underway in spring 2022, funded by a one-million-dollar grant from the US Department of Agriculture.

The identities of participating ranchers have been kept secret—a reflection of the polarized nature of the work—but Nickerson confirms that the team is indeed working with ranchers in Washington State. This sort of collaboration gives some environmental groups heartburn, making them fret that the anti-wolf caucus will use the study as a weapon. Nickerson acknowledges these concerns, but, much like Daniel Curry, she also insists on the importance of learning from the people who live in wolf country about what it means to live there. To that end, researchers have shared their proposed methodology with the ranchers participating in the study and solicited feedback from them. Researchers will also be collecting rancher's and landowner's observations and opinions about wolves.

"In the scientific community," Nickerson tells me, "folks don't like to talk about 'observation as proof.' But in this project, we're trying to take that seriously. We know that the livestock producers have generational knowledge, and they know what to look for on the landscape."

As an undergraduate, Nickerson became interested in this line of study after a professor told her, "It's not this or that. Life is about this *and* that." That ethos has extended into her academic interests, and she sees the return of predators to the landscape as a perfect example of "this and that."

That broad, multidisciplinary approach is reflected in the scope and scale of the study, even if the primary question is rather simple. "How much does it cost to put these practices on the

landscape and how effective are these practices at reducing predation?" wonders Alex Few, a coordinator for the Western Landowners Alliance, one of the organizations participating in the study.

Results are still years away, but the effort reveals a growing rec-
ognition that when it comes to wolves and cattle, focusing on the
hyperlocal will be critical. "The more engaged we can be in relationship with the world around us," says Nickerson, "the closer
we can get to healthy, happy landscapes."

On a cloudy and damp fall day, I drive to Daniel Curry's home for
the first time. The chill of oncoming winter is already in the air.
The land has grown quiet, as if preparing itself for what's ahead.
Visiting Curry at home is a big deal, and I'm nervous. He's painfully private, constantly questioning whether he wants to keep
talking to a reporter about some of the more intimate aspects of
his life and work.

But he's invited me up to see his house and so I go.

He lives about forty minutes north of the nearest town, not far
from the Canadian border. On his sixty-three acres of land he has
a cave, a stream, and meadows. The log home was built in stages,
starting in 1970. It's quiet inside, the occasional noise from a passing car muffled by a thick wall of pines, larch, and aspens. Curry's
home is spartan, tidy to the point of emptiness.

The land outside is more chaotic, peppered with pastures and
vehicles. Curry shows me his sailboat, Bumblebee, parked near
a large barn. His three black Dobermans—Rook, Knight, and
Bishop—rocket around, careening off anything dumb enough to
get in their way. Curry tells me that he's seen wolf tracks near his
home, in addition to the regular moose, deer, elk, and bear sign.

On one occasion he saw a wolf on his property. He's raised goats, sheep, and horses.

From his house we head uphill, following an old logging road. We pass an old beaver dam, the original residents long ago trapped by a neighbor. Next is a small pet cemetery where Curry has buried his animals that have died. Soon we come to a junction where the road splits, and we head to the right, leveling out on the side of a hill.

Like all landowners I've met, Curry talks with evident pride and care for his land, telling stories and sharing his dreams, recounting the history of how he came to be there.

He bought the property in 2014 after spending nearly seven months alternating between living on public land and staying with friends. This had been a tough period to get through, but it did help him cement his reputation as a dedicated range rider.

"I had no home to go back to and sit down and watch a movie and drink a beer," he says of that time. "So I just plowed my nose into [range riding]."

When the house went on the market, a friend gave Curry a heads-up. He soon discovered that the owner, Tony Bolles, had sponsored a wolf at Wolf Haven. A bizarre coincidence, but things would get stranger.

Bolles told Curry he'd bought the house in 1991 from the man who'd built it twenty-one years earlier. A man named William White.

The Whites are infamous in the wolf world. That's because in December 2008, a FedEx worker discovered that blood was oozing from a package addressed to Canada. It turned out to be a wolf pelt. White's son Tom had killed the animal and was mailing it to a friend in Alberta. For his part in the crime, the elder White eventually pleaded guilty to conspiracy to kill an endangered animal. Tom White pleaded guilty for the killing, and his wife Erin, who had mailed the package, pleaded guilty to conspiracy and unlawfully exporting an endangered species. Curry knew all about the case, and had been apoplectic when he first heard the story. He went so far as to look up aerial photos of White's home and concoct a vague plan, which he didn't act upon, to confront him.

Six years later, he ended up buying a home built by the man he'd once surveilled. The whole thing seemed impossible.

And then William White called. He wanted to see the place he'd built forty-odd years ago. A bit stunned by this turn of events, Curry agreed and soon found himself sitting across from William White.

"It was weird, because you're like some wolf-poaching piece of crap, and I'm some wolf-loving piece of crap in your mind, and we're sitting here," he says of the meeting. "If you take all that crap away, I liked the guy a lot."

Despite yearly losses to poachers, to the state, and due to car accidents, the wolf population in Washington has grown every year. Wolves are not single-handedly destroying ranching or endangering humans—in fact, Washington's ranching industry was already besieged by larger market pressures long before wolves arrived.

But Curry believes that tolerance for wolves is on the ropes. The rhetoric has turned increasingly violent. Every time a wolf kills a cow or the state kills a wolf, the divide widens, and meetings like the one between Curry and White become less likely.

———

And yet, a shiver of fate runs throughout Curry's life. How else can he explain the coincidence of him ending up in William White's home? Or that feeling he gets—the bone-deep familiarity and calmness—when he hears a wolf howling?

He can't. But he does remember a dream he started having when he was four years old, the year before he dedicated his life to the animal world.

"I was sitting across the street from this campfire with this girl," he tells me. "We were sitting there talking when a wolf came up and attacked me and pulled me away. That was the only recurring dream I've ever had in my life."

It occurs to me that the broad strokes of Curry's dream align with the foundational lore of the Quileute tribe in western Washington. According to legend, their people were born when a wandering spirit transformed wolves into humans. And every year, the tribe reenacts the story of a young man kidnapped by wolves. Failing to kill him, they instead teach him their ways, eventually returning him to his own people.

When I ask Curry if he's familiar with the story, he's stunned, never having heard it before. And so I share the following passage with him from Barry Lopez's book *Of Wolves and Men*:

> The wolves tried to kill him but could not and so they became his friends. They taught him about themselves, then sent him back to his village to teach his tribe the rites of the wolf ceremony. The young man told his people that it was necessary for the strength of the tribe, for their success in war, and everything else they did, that they should be like

wolves. They must be as fierce, as brave, and as determined as the one who is the greatest hunter in the woods. In this ceremony people are "stolen" by wolves, go through a terrifying confrontation, and emerge wolflike.

In a culture defined by reductive thinking, it's easy to discount stories and myths, to discount observational knowledge, generational wisdom. We like to pretend that complex things can always be broken down into their simplest components.

Consider the widening gulf at the heart of the wolf debate. Many wolf lovers ignore or denigrate the ranchers and farmers who hold opposing views. Meanwhile, many farmers ignore the wolves, or the environmentalists who love them. Shouting facts at one another does not bridge that divide. Nevertheless, as suggested by Curry's dream, and by the encoded generational wisdom of the Quileute myth, to learn how to live with the other—be it wolf or human—always requires a sacrifice.

Daniel Curry has offered that sacrifice. He has given his life over to the animal other, and he's paid a price. Sometimes he overstates things, but in this he is right: the Wolf Wars are a symptom of a bigger problem, a rift that's fractured our politics and culture. Striking a balance in wolf land would go a long way toward kneading the dough of society back together.

ACKNOWLEDGMENTS

To Tyler Tjomsland, whose skill and insight as a reporter and photographer is matched only by his appetite for Taco Bell. Our reporting adventures keep me in the industry. Gratitude to Kevin Taylor for years of encouragement and advice, long before I had earned either. To Ben Goldfarb for helping point the way forward and introducing me to this book's ultimate home, Timber Press. Thanks to Michael Heald and Will McKay for lifesaving edits. To the *Spokesman-Review* for providing the time and space to work on this project. Finally, thanks to Dr. Valerius Geist for his wit and wisdom. Dr. Geist passed away on July 6, 2021.

NOTES AND FURTHER READING

INTRODUCTION

Arksey, Laura. "Pend Oreille County—Thumbnail History." *HistoryLink.org*, January 22, 2006. https://www.historylink.org/file/7618.

National Park Service. "The Greater Yellowstone Ecosystem." Last modified August 21, 2020. https://www.nps.gov/yell/learn/nature/greater-yellowstone-ecosystem.htm#:~:text=The%20Greater%20Yellowstone%20Ecosystem.,temperate%2Dzone%20ecosystems%20on%20Earth.

Walters, Daniel. "Rep. Shea's secret group discussed sending severed wolf tail and testicles to environmental activist." *Inlander*, May 7, 2017. https://www.inlander.com/spokane/rep-sheas-secret-group-discussed-sending-severed-wolf-tail-and-testicles-to-environmental-activist/Content?oid=17572893.

Schwarz, D. Hunter. "The fastest growing states in the US are all out West." *Deseret News*, December 27, 2021. https://www.deseret.com/2021/12/27/22855777/the-fastest-growing-states-in-the-u-s-are-all-out-west-utah-idaho-arizona-montana.

Easter, Michael. *The Comfort Crisis: Embrace Discomfort to Reclaim Your Wild, Happy, Healthy Self.* New York: Rodale Books, 2021.

Sterba, Jim. *Nature Wars: The Incredible Story of How Wildlife Comebacks Turned Backyards into Battlegrounds*. New York: Broadway Books, 2012.

CHAPTER 1: THE WOLF

United States Census Bureau. "QuickFacts, Pend Oreille County, Washington." https://census.gov/quickfacts/fact/table/ pendoreillecountywashington,WA/PST045219.

Some of the history of wolves in Washington in the 1950s comes from contemporaneous press accounts from the *Spokesman-Review* archives.

Abram, David. *The Spell of the Sensuous: Perception and Language in a More-Than-Human World*. New York: Vintage Books, a division of Penguin Random House LLC, 2017.

Mech, L. David. *The Wolf: The Ecology and Behavior of an Endangered Species*. New York: Natural History Press, 1970.

Leonard, Jennifer A., et al. "Megafaunal Extinctions and the Disappearance of a Specialized Wolf Ecomorph." *Current Biology* 17, no. 13 (July 2007).

Flores, Dan L. *Coyote America: A Natural and Supernatural History*. New York: Basic Books, 2016.

CHAPTER 2: CRYING WOLF

Lopez, Barry Holstun. *Of Wolves and Men*. New York: Scribner, 2004.

Blakeslee, Nate. *American Wolf: A True Story of Survival and Obsession in the West*. New York: Broadway Books, 2018.

Coppinger, Raymond, and Lorna Coppinger. *Dogs: A Startling New Understanding of Canine Origin, Behavior, and Evolution*. Romford: Crosskeys Select Books, 2004.

Curry, Daniel. Personal journals, 2008.

Curry, Daniel. "A boy named Sioux." *Wolf Tracks*, summer 2010. https://www.wolfhaven.org/wp-content/uploads/2019/01/Sioux.pdf.

Curry, Daniel. "Our Brita." *Wolf Tracks*, summer 2008. https://www.wolfhaven.org/wp-content/uploads/2019/01/Brita.pdf.

Mann, Charles C. *1491: New Revelations of the Americas before Columbus.* New York: Alfred A. Knopf, 2012.

Flores, Dan. "Essay: The Great Plains 'Wilderness' as a Human-Shaped Environment." *Great Plains Research* 9, no. 2 (Fall 1999).

McIntyre, Rick, ed. *War Against the Wolf: America's Campaign to Exterminate the Wolf.* St. Paul: Voyageur Press, 1995.

Corbin, Benjamin. *Corbin's Advice, or the Wolf Hunter's Guide: Tells How to Catch 'Em and All About the Science of Wolf Hunting.* Bismark: The Tribune Co., 1900.

Stuart, Granville. *Forty Years on the Frontier.* Lincoln: University of Nebraska Press, 1977.

Paige M. Nash. "The Salmon Chief." *Spokane Historical,* accessed May 31, 2022. https://www.spokanehistorical.org/items/show/508.

Kalispel National Resources Department. "About KNRD." https://www.knrd.org/about/.

International Wolf Center. "Hunting and Feeding Behavior." https://www.wolf.org/wolf-info/basic-wolf-info/biology-and-behavior/hunting-feeding-behavior/.

Washington Department of Fish and Wildlife et al. "Washington Gray Wolf Conservation and Management 2021 Annual Report." Washington Department of Fish and Wildlife, April 9, 2022. https://www.wdfw.wa.gov/sites/default/files/publications/02317/wdfwo2317.pdf.

Struhsaker, Peggy. "Wolves in the Northeast: Principles, Problems and Prospects." National Wildlife Federation, 2003. https://www.nwf.org/~/media/pdfs/wildlife/wolvesnortheast.pdf.

Toffler, Alvin. *Future Shock.* New York: Ballantine Books, 2022.

Mech, L. David. "Is Science in Danger of Sanctifying the Wolf?" *Biological Conservation* 150, no. 1 (2012): 143–149. https://doi.org/10.1016/j.biocon.2012.03.003.

CHAPTER 3: LOOKING FOR SUPPORT

Bargreen, Melinda. "Classical music comes home to the 'It Girl's' house." *The Seattle Times,* March 8, 2007. https://www.seattletimes.com/entertainment/classical-music-comes-home-to-the-it-girls-house/.

Moskowitz, David. *Wolves in the Land of Salmon*. Portland: Timber Press, 2013.

Wiles, Tay, and Brooke Warren. "Federal-lands ranching: A half-century of decline." *High Country News*, June 13, 2016. https://www.hcn.org/issues/48.10/federal-lands-grazing.

CHAPTER 4: INTO THE WOODS, INTO THE WEEDS

Maletzke, Benjamin T., et al. "A meta-population model to predict occurrence and recovery of wolves." *The Journal of Wildlife Management* 80, no. 2 (2015): 368–376. https://doi.org/10.1002/jwmg.1008.

Maletzke, Ben, and Dan Brinson. "2020 Annual Wolf Report." The Washington Department of Fish and Wildlife, 2020. https://www.wdfw.wa.gov/sites/default/files/2021-04/5wolfreportpp.pdf.

Tate, Cassandra. "Kettle Falls." *HistoryLink.org*, December 27, 2005. https://www.historylink.org/file/7577.

McIntosh, Robert P. "The Background and Some Current Problems of Theoretical Ecology." *Conceptual Issues in Ecology* (1982): 1–61. https://doi.org/10.1007/978-94-009-7796-9_1.

Quammen, David. *The Song of the Dodo: Island Biogeography in an Age of Extinctions*. New York: Scribner, 2004.

MacArthur, Robert H., and Edward O. Wilson. *The Theory of Island Biogeography*. Princeton: Princeton University Press, 2016.

Lockwood, Dale. "When Logic Fails Ecology." *The Quarterly Review of Biology* 83, no. 1 (2008): 57–64. https://doi.org/10.1086/529563.

CHAPTER 5: THE CATTLE

Tri-State Livestock News. "Country of Origin Labeling: MCOOL bill officially introduced." *Tri-State Livestock News*, September 24, 2021. https://www.tsln.com/news/country-of-origin-labeling-mcool-bill-officially-introduced/.

Miles, Tom, and Rod Nickel. "WTO rules against U.S. in meat labeling case." *Reuters*, June 29, 2012. https://www.reuters.com/

article/us-trade-usa-canada-mexico/wto-rules-against-u-s-in-meat-labeling-case-idUSBRE85S0Y920120629.

Lynch, David J. "'America First' may be last hope for these cattle ranchers." *The Washington Post*, May 3, 2019. https://www.washington post.com/business/economy/america-first-may-be-last-hope-for-these-cattle-ranchers/2019/05/03/7469d1de-5bad-11e9-9625-01d48d50ef75_story.html.

Laporte, Isabelle, et al. "Effects of Wolves on Elk and Cattle Behaviors: Implications for Livestock Production and Wolf Conservation." *PLOS ONE* 5, no. 8 (2010). https://doi.org/10.1371/journal.pone.0011954.

van der Voo, Lee. "Betting the Ranch." *High Country News*, December 1, 2021. https://www.hcn.org/issues/53.12/ranching-betting-the-ranch.

MacDonald, James M., and Robert A. Hoppe. "Examining Consolidation in U.S. Agriculture." *USDA Economic Research Service*, March 14, 2018. https://www.ers.usda.gov/amber-waves/2018/march/examining-consolidation-in-us-agriculture/.

CHAPTER 6: PREDATORS AND PREY

Washington Predator-Prey Project. https://predatorpreyproject.weebly.com.

Dellinger, J. A., et al. "Impacts of Recolonizing Gray Wolves (Canis Lupus) on Survival and Mortality in Two Sympatric Ungulates." *Canadian Journal of Zoology* 96, no. 7 (2018): 760–768. https://doi.org/10.1139/cjz-2017-0282.

Goldfarb, Ben. *Eager: The Surprising, Secret Life of Beavers and Why They Matter*. White River Junction: Chelsea Green Publishing, 2018.

Brice, Elaine M., et al. "Sampling Bias Exaggerates a Textbook Example of a Trophic Cascade." *Ecology Letters* 25, no. 1 (January 2022): 177–188. https://doi.org/10.1101/2020.05.05.079459.

Klauder, Kaija J., et al. "Gifts of an enemy: scavenging dynamics in the presence of wolves (Canis lupus)." *Journal of Mammalogy* 102, no. 2 (April 2021): 558–573. https://doi.org/10.1093/jmammal/gyab020.

Beausoleil, Richard A., et al. "Long-Term Evaluation of Cougar Density and Application of Risk Analysis for Harvest Management." *The Journal of Wildlife Management* 85, no. 3 (2021): 462–473. https://doi.org/10.1002/jwmg.22007.

Peakbagger.com. "Most Remote Spots in USA Wilderness Complexes." https://www.peakbagger.com/report/report.aspx?r=w.

USDA Forest Service. "Road Management Website." https://www.fs.fed.us/eng/road_mgt/overview.shtml.

CHAPTER 7: PREDATORS AND HUMANS

Miller, Craig A., and Jerry J. Vaske. "How State Agencies Are Managing Chronic Wasting Disease." *Human Dimensions of Wildlife* (2022): 1–10. https://doi.org/10.1080/10871209.2021.2023712.

Meeks, Abigail, et al. "Hunter Concerns and Intention to Hunt in Forested Areas Affected by Wildlife Disease." *Forest Science* 68, no. 1 (February 2022): 85–94. https://doi.org/10.1093/forsci/fxab049.

Quammen, David. *Spillover: Animal Infections and the Next Human Pandemic*. New York: W. W. Norton & Company, 2012.

Randolph, Delia Grace, et al. "Preventing the Next Pandemic: Zoonotic diseases and how to break the chain of transmission." Nairobi, Kenya: United Nations Environment Programme and International Livestock Research Institute, 2020.

Loh, Elizabeth H., et al. "Targeting Transmission Pathways for Emerging Zoonotic Disease Surveillance and Control." *Vector-Borne and Zoonotic Diseases* 15, no. 7 (July 2015): 432–437. https://doi.org/10.1089/vbz.2013.1563.

Joselow, Maxine, and Alexandra Ellerbeck. "Biden Is Approving More Oil and Gas Drilling Permits on Public Lands than Trump, Analysis Finds." *The Washington Post*, December 6, 2021. https://www.washingtonpost.com/politics/2021/12/06/biden-is-approving-more-oil-gas-drilling-permits-public-lands-than-trump-analysis-finds/.

CHAPTER 8: WHAT OF THE WOLF?

Santiago-Ávila, Francisco, and Adrian Treves. "Killing Wolves to Prevent Predation on Livestock May Protect One Farm but Harm Neighbors: Variables and Sample STATA Code for Survival Analytics." *Protocols* (2017). https://doi.org/10.17504/protocols.io.j2rcqd6.

de Waal, Frans. *Are We Smart Enough to Know How Smart Animals Are?* New York: W. W. Norton & Company, 2017.

Treves, A., et al. "Just Preservation." *Biological Conservation* 229 (January 2019): 134–141. https://doi.org/10.1016/j.biocon.2018.11.018.

Liberg, Olof, et al. "Shoot, Shovel and Shut up: Cryptic Poaching Slows Restoration of a Large Carnivore in Europe." *Proceedings of the Royal Society B: Biological Sciences* 279, no. 1730, (March 7, 2012): 910–915. https://doi.org/10.1098/rspb.2011.1275.

Linnell, John D. C., and Julien Alleau. "Predators That Kill Humans: Myth, Reality, Context and the Politics of Wolf Attacks on People." *Problematic Wildlife* (2015): 357–371. https://doi.org/10.1007/978-3-319-22246-2_17.

Linnell, John D. C., et al. "Wolf attacks on humans: an update for 2002–2020." *Norwegian Institute for Nature Research* 1944 (2021). https://cupdf.com/document/wolf-attacks-on-humans-an-update-for-20022020.html?page=1.

Linnell, John D. C., et al. "The Fear of Wolves: A Review of Wolf Attacks on Humans." *Norwegian Institute for Nature Research* 731 (2002). https://www.researchgate.net/publication/236330045_The_fear_of_wolves_A_review_of_wolf_attacks_on_humans.

Geist, Valerius. "Human use of wildlife and landscapes in pre-contact southern North America, as recorded by Alvar Nunez Cabeza de Vaca 1527–1536." *Beiträge zur Jagd-und Wildforschung* 43 (2018): 397–406.

Geist, Valerius. "Wolves, Bears and Human Anti-Predator Adaptations." Unpublished.

Graves, Will N., and Valerius Geist, ed. *Wolves in Russia: Anxiety through the Ages*. Calgary: Detselig Enterprises, 2007.

Marshall, Andy. "Wildlife Scientist Devotes Rich Life to Educating the Public." *Wild Lands Advocate* 12, no. 5 (2004). https://www.albertawilderness.ca/wp-content/uploads/2015/0 9/2004-valerius-geist_20041001_WLA_AR_VG.pdf.

Geist, Valerius. "Wolves: When Ignorance Is Bliss." *Idaho for Wildlife*. https://www.Idahoforwildlife.com/Website%20articles/Dr%20Geist/ When%20ignorance%20is%20obliss.html.

MacKinnon, J. B. "Death of a Modern Wolf." *Hakai Magazine*, October 17, 2017. https://www.hakaimagazine.com/features/death-modern-wolf/.

Flores, Dan. "Essay: The Great Plains 'Wilderness' as a Human-Shaped Environment." *Great Plains Research* 9, no. 2 (Fall 1999).

CHAPTER 9: WOLF POLITICS

Nark, Jason. "When Wolves Made a Resurgence, Her Job Was to Make Peace Between Ranchers and Conservationists." *The Washington Post*, December 5, 2018. https://www.washingtonpost.com/ lifestyle/magazine/when-wolves-made-a-resurgence-her-job-was-to-make-peace-between-ranchers-and-conservationists/ 2018/12/03/fb825dca-e78c-11e8-bbdb-72fdbf9d4fed_story.html.

Read, Richard. "One Ranch, 26 Wolves Killed: Fight over Endangered Predators Divides Ranchers and Conservationists." *Los Angeles Times*, December 18, 2019. https://www.latimes.com/world-nation/ story/2019-12-18/endangered-wolf-killings-ranch.

Scott, Aaron, host. *Timber Wars*. Podcast audio. Oregon Public Broadcasting, 2020. https://www.opb.org/show/timberwars/.

Abbey, Edward. *The Monkey Wrench Gang*. New York: HarperCollins Publishers, 2006.

Landers, Rich. "Collaborative Effort: A Washington Cattleman and Biologist Are Working to Reduce Wolf-Livestock Conflicts." *The Spokesman-Review*, April 2, 2018. https://www.spokesman.com/ stories/2018/mar/29/collaborative-effort-a-washington-cattlemen-and-bi/.

Solomon, Christopher. "Who's Afraid of the Big Bad Wolf Scientist?" *The New York Times*, July 5, 2018. https://www.nytimes.com/2018/07/05/magazine/whos-afraid-of-the-big-bad-wolf-scientist.html.

Fischer, Hank. *Wolf Wars: The Remarkable Inside Story of the Restoration of Wolves to Yellowstone*. Helena: Falcon Press, 2003.

Beissinger, Steven R. "The North American Model of Wildlife Conservation. Wildlife Management and Conservation. Edited by Shane P. Mahoney and Valerius Geist." Book review. *The Quarterly Review of Biology* 96, no. 2, (2021): 153. https://doi.org/10.1086/714448.

Manfredo, Michael J., et al. "Values, Trust, and Cultural Backlash in Conservation Governance: The Case of Wildlife Management in the United States." *Biological Conservation* 214 (October 2017): 303–311. https://doi.org/10.1016/j.biocon.2017.07.032.

Mapes, Linda. "A War Over Wolves." Seattle Times, 2017. https://projects.seattletimes.com/2017/wsu-wolf-researcher-wielgus.

Organ, J. F., et al. *The North American Model of Wildlife Conservation*. Bethesda: The Wildlife Society, 2012.

CHAPTER 10: ANOTHER WAY

Grandin, Temple. "Principles for Low Stress Cattle Handling." *Proceedings, the Range Beef Cow Symposium* XVI (December 1999): 134. https://digitalcommons.unl.edu/rangebeefcowsymp/134/.

Louchouarn, Naomi X., and Adrian Treves. "Low-Stress Livestock Handling Protects Cattle in a Five-Predator Habitat." Preprint, 2021. https://doi.org/10.21203/rs.3.rs-1061804/v1.

Goldfarb, Ben. "How Should We Treat Fish Before They End Up on Our Plates?" *High Country News*, March 20, 2019. https://www.hcn.org/issues/51.6/fish-how-should-we-treat-fish-before-they-end-up-on-our-plates.

Turner, Jack. *The Abstract Wild*. Tucson: University of Arizona Press, 1999.

Lopez, Barry Holstun. *Of Wolves and Men*. New York: Scribner, 2004.

INDEX